Diagnostic Genetic Te

David Bourn

Diagnostic Genetic Testing

Core Concepts and the Wider Context for Human DNA Analysis

 Springer

David Bourn
Northern Genetics Service
Newcastle upon Tyne, UK

ISBN 978-3-030-85509-3   ISBN 978-3-030-85510-9   (eBook)
https://doi.org/10.1007/978-3-030-85510-9

This Springer imprint is published by the registered company Springer Nature Switzerland AG
The registered company address is: Gewerbestrasse 11, 6330 Cham, Switzerland

For Claire, Rachel and Stephen (in alphabetical and inverse height order)

Foreword

Genetic testing has been used by healthcare services since the 1960s when it was initially confined to looking for chromosome abnormalities down the microscope. DNA studies came to the fore in the 1980s, at first offering indirect testing by genetic linkage analyses, and subsequently direct analyses by DNA sequencing of disease-associated genes. The Human Genome Project (HGP), a huge international endeavour that began in 1990 and took 13 years, at a cost of $3 billion, has enormously extended the possibilities for genetic testing. By providing a genome-wide human DNA reference sequence, it permitted for the first time genome-wide studies that we now call genomics.

Now let's fast forward to the present day. What has changed? In a word, everything. DNA sequencing technology has advanced so rapidly that what took the HGP 13 years and cost $3 billion can be achieved today in just one day and for a few hundred dollars only. The era of personal genome sequencing, begun in 2007, is now in full flight and accelerating. Current estimates indicate that of the order of 60 to 160 million humans will have their genomes sequenced by 2035. Plans for neonatal genome screening have also been considered in some countries and may be implemented following public consultation.

As genomes become readily available for analysis, so the scope of genomics and the nature of genetic testing are being transformed. Genome sequencing is beginning to transition to routine health care and national genomic medicine services have begun, or been planned, in several advanced countries.

An agenda of 'mainstreaming genetics' envisages that any branch of healthcare practice should be sufficiently versed in the implications of particular genetic tests to offer such tests directly to their patients.

With any technological leap, comes not just hope, but challenges. Take sequence interpretation. In genetic testing, comparison of the DNA from the person being tested against a control reference human DNA sequence helps to identify a damaging DNA variant associated with disease. That can sometimes be difficult at the gene level, but at the genome level, there may be millions of changes in the test sample when compared with the reference sequence. Some of the changes may be difficult to interpret; others may turn out to be ones associated with a completely different disease than the one for which the test was ordered.

Direct-to-consumer genetic testing services are also now being offered by different companies, bypassing healthcare services altogether. Ethical issues are important in all types of genetic testing (because unlike other types of medical test, genetic tests may have consequences for close relatives in addition to the person being tested), but can be expected to be especially so in genome-wide testing.

September 2021

Prof. Tom Strachan
FRSE FMedSci
Newcastle University
Newcastle upon Tyne, UK

Preface

The primary aim of this book is to introduce some of the most important concepts underlying genetic testing, in families with inherited conditions and in patients with cancer. Specific examples of gene testing have been chosen with a view to illustrating particular themes. Many other examples could have been used; this is by no means an exhaustive treatment of diagnostic genetics, but hopefully the key details are still captured. The secondary aim is to provide some broader context, going beyond details of the laboratory work. In addition to the immediate implications for the patients being tested and their families, there are wider consequences to be considered. The specific scientific content on genes and genetics is largely uncontroversial (where there are uncertainties this should be explicitly stated). The division between this reliable information and my own views should be clear, and in this sense, the book can be said to be both good and original—but as Samuel Johnson (almost) put it, the good parts are not always original and the original parts not necessarily good.

Any technical terms used are defined either in the main body of the book or in the additional material provided in the footnotes. Occasional boxes scattered through the chapters are used to convey more complex information, or to supply material that extends the themes covered in the main text. Key laboratory methods used in DNA analysis are introduced, but rather than listing extensive technical details, the emphasis is on an attempt to provide a broad understanding of how and why these methods are used. Detailed information can be accessed via the selection of reliable and well-curated websites listed below. Together these should provide links to a comprehensive body of resources covering the scientific basis of the techniques used for genetic analysis, specific details on genetic disorders, general discussions on the role of genetic testing in healthcare, and support groups for patients and their families.

Further Reading

Laboratory Techniques

The Lab Tests Online site provides an overview of the methods used with copious links to more detailed information. There are many other websites going into as much detail on specific techniques as could possibly be required, frequently including animations that help to make sense of the more complex methods.
https://labtestsonline.org/genetic-testing-techniques

Genetic Disorders

A wealth of up to date information on genetic disorders can be found at the GeneReviews and Online Mendelian Inheritance in Man (OMIM) websites (https://www.ncbi.nlm.nih.gov/books/NBK1116/ and https://www.omim.org/ respectively). These provide a way in to the primary literature where more detail is of interest.

General Overviews of Genomic Testing in Healthcare

Of many possibilities, two are given below.
Health Education England Genomics Education Programme
https://www.genomicseducation.hee.nhs.uk/education/core-concepts/what-is-genomics/
National Human Genome Research Institute Genomics Education Websites
https://www.genome.gov/about-genomics/teaching-tools/Genomics-Education-Websites

Patient Support Groups

There are numerous groups providing support for patients with a genetic condition and for their families, which can easily be found online by searching for the disorders in question. Collated information about support groups and many other resources for patients and their families can be found at the Genetic Alliance website.
http://www.geneticalliance.org/

Newcastle upon Tyne, UK   David Bourn

Acknowledgements

Many thanks to Gillian Bourn, who read and corrected the entire book (so many misplaced commas) and introduced multiple improvements. Many thanks also to Rachel Bourn for perceptive and helpful comment on all of the chapters, and to Nick Bown and Ciaron McAnulty who both commented on specific chapters. Tanja Weyandt at Springer was supportive and helpful throughout the development of this book, and suggested a number of beneficial changes to the format. Specific thanks to those who provided figures as noted in the captions (and apologies if I have missed anyone out) and thanks to Stephen Bourn for help with the cover image. I've learned from many colleagues in research and diagnostic genetics over the years, but especially from Ann Curtis, John Goddard, Fiona Macdonald, Roger Mountford and Tom Strachan.

All of the author's royalties will go to the Newcastle Hospitals Charity.

Contents

About the Author

David Bourn, Ph.D., FRCPath is a UK-based state registered clinical scientist and the head of an NHS Regional Genetics laboratory, who has worked in the NHS for more than 25 years. Prior to his involvement in diagnostic genetics, he spent some years in research, with his Ph.D. and the bulk of his postdoctoral work in the field of human molecular genetics. His links with Human Genetics Departments go back to the 1980s and in addition to his familiarity with the evolution of genetic testing over the past three decades, he has an enduring interest in the wider implications of this testing.

Abbreviations

AIS	Androgen insensitivity syndrome
ARMS	Amplification refractory mutation system
AS	Angelman syndrome
BMD	Becker muscular dystrophy
BMT	Bone marrow transplant
cDNA	Complementary DNA
CF	Cystic fibrosis
CML	Chronic myeloid leukaemia
ctDNA	Circulating tumor DNA
DMD	Duchenne muscular dystrophy
DNA	Deoxyribonucleic acid
DTC	Direct-to-consumer
FH	Familial hypercholesterolaemia
FISH	Fluorescence in situ hybridization
FXPOI	Fragile X-associated premature ovarian insufficiency
FXTAS	Fragile X associated tremor/ataxia syndrome
HD	Huntington disease
MCC	Maternal cell contamination
MLPA	Multiplex ligation-dependent probe amplification
mRNA	Messenger RNA
NGS	Next-generation sequencing
PCR	Polymerase chain reaction
PGD	Pre-implantation genetic diagnosis
PND	Prenatal diagnosis
PWS	Prader-Willi syndrome

RNA	Ribonucleic acid
SBMA	Spinal and bulbar muscular atrophy
STR	Short tandem repeat
UPD	Uniparental disomy
VUS	Variant of uncertain significance
WES	Whole exome sequencing
WGS	Whole genome sequencing

1

Genetic Testing, Some Themes and Some Basics

Summary This chapter provides a brief overview of genetic testing for inherited diseases and cancer. Some of the recurrent themes of the book are introduced, including complexity, genetic determinism, aspects of risk and uncertainty, and the value of understanding biological characteristics in terms of a continuum, rather than attempting artificial categorisation. Powerful current developments in diagnostic genetics are described briefly, together with a suggestion of the limitations that remain. The chapter includes selected basics of genetics, covering human genome architecture, DNA as an information store, the processes involved in the flow of that information into a functional gene product, and the mutations that disrupt the flow. Patterns of inheritance (autosomal dominant, autosomal recessive, X-linked and mitochondrial) are introduced. The final part of the chapter gives a succinct description of some of the basic methodology of diagnostic DNA analysis, including microarrays, the polymerase chain reaction, electrophoresis and DNA sequencing.

Genetic Testing

Being given the result of a genetic test can force the patient to confront an unforgiving dichotomy, an inescapable fork in the road. Either you did inherit a damaging gene variant or you didn't, and if you did, then either you have or have not passed that variant on to a child. In terms of cancer, your tumor either does or does not carry a genetic change that permits a specific treatment or confers a poor prognosis. Apparently no ambiguity; no middle ground.

Of course, the same can be said for the results of many medical tests, but there are often wider implications for DNA testing.[1] Finding an error in your genes cuts right to the heart of your identity and in addition to the direct consequences for your existing family members, who may be at risk of suffering or passing on the same condition, there are implications for future generations. Over the last 30 years or so the scope of genetic testing has grown exponentially, driven by increases in knowledge and perhaps even more by advances in the available technology. Although genetic diseases are individually rare, there are thousands of inherited conditions and cumulatively it is estimated that rare genetic diseases (defined as affecting fewer than 1 in 2000 of the population) may be present in as many as 1 in 15 of us. More and more of these conditions can be tested (although whether actual treatments for patients have kept pace is a different question). DNA testing has the huge advantage that essentially the same generic technologies can be applied to all of these diseases, without having to design a unique test based on the underlying biological cause of each separate condition. Beyond inherited disease, genetic testing is becoming increasingly important in the diagnosis and treatment of cancer (effectively all cancers start with damage to the DNA in a single cell).

During the last three decades of rapid expansion in the scientific repertoire available to diagnostic genetics, there has been a corresponding increase in the number of medical professionals working in the field, aiming to provide accurate and timely results to help in the diagnosis of genetic disease and cancer. Genetic testing may have great predictive power, although perhaps less than is often supposed (a theme that will crop up frequently in this book). The implications beyond the individual being tested, and the possibility of diagnosing future disease, give rise to ethical problems unique to the field. Testing even for a single gene condition can have unexpected consequences depending on the context, and the scope of the testing that is now routinely carried out can be so broad that unexpected, uncertain and unwanted findings are almost guaranteed. In addition to introducing the central aspect of diagnostic genetic testing, finding errors in the DNA sequence of patients that might explain their clinical symptoms, I also aim to explore some of

---

[1] DNA (deoxyribonucleic acid in full) stores the genetic information for all complex life and is used to transmit this information through the generations. The related molecule RNA (ribonucleic acid) is used as an intermediary in the processes involved in moving from the coded information in DNA to the proteins forming the complex structures and facilitating the chemical reactions that make up living organisms. RNAs are also functional molecules in their own right, performing a range of roles in the cell.

the wider and less obvious ramifications of DNA testing. Allusions to genes, inheritance and DNA are increasingly part of the general conversation.

Complexity: Genes and Environment

At first sight DNA testing seems to give clear-cut answers—a change in the DNA is digital, either present or not. This description of the binary impact of a DNA test does hold good (with some caveats) for the majority of the single gene conditions tested for routinely in genetic laboratories. However, for most genetic variants the situation is less straightforward. Rather than a change in an individual's DNA inevitably leading directly to a clinical problem or a change in some other characteristic—genetic determinism—the outcome is in most cases affected by a myriad of other factors, internal and external to that individual. The idea of genetic determinism is off-putting for many who dislike the idea of humans as puppets, with their genes pulling the strings, but in practice almost everyone would acknowledge that environmental and cultural influences play a major role. There is a spectrum of human characteristics with, at one extreme, features entirely determined by a particular gene and, at the other, features that are entirely environmental. In most cases, both genes and environment play a part, and for any complex aspects of structure or behaviour many genes may be involved, interacting with each other and the environment in intricate networks. Setting up a competition between nature and nurture, genes and environment, is rarely the best way to frame the discussion. Take the example of language: in one sense this is entirely genetic, in that the capacity for complex communication, on this planet at least, seems to be restricted to humans. However, the specific language you speak will be wholly determined by your environment—no one has a fully formed private language.

Complexity is the rule rather than the exception. Even where errors in a single gene are largely responsible for causing a particular disorder, there will almost always be modifying factors, including the effects of other genes as well as external environmental factors. The consequences of variations in the DNA sequence are therefore not always completely predictable, a theme which will be explored through examining the effects of deleterious changes in some of the genes involved in genetic disease.

Risk and Uncertainty

An alternative perspective to that of genetic determinism encompasses chance (using chance in the sense of probability rather than randomness). The concepts of risk and uncertainty are of fundamental importance in the context of genetic testing. In some cases, the risks are clear-cut, where inheritance patterns for genetic disorders follow straightforward rules (see below for a brief description of dominant, recessive and X-linked inheritance). However, not all disorders follow classic dominant or recessive patterns. The uncertainties of some genetic results, whether caused by intrinsic complexities of the genetics, environmental factors, or just the current state of ignorance regarding the implications of a particular finding, can be hard to convey to patients and their families. It can be difficult to express probabilities in a meaningful way and, given that perceptions of risk differ widely and may depend on how the risk is framed, this is one of the areas that make genetic counselling (explaining to patients the often complex implications of the result of a genetic test) such a demanding task.

DNA and Categorisation

Scientific (or often pseudoscientific) language is indiscriminately employed in debates around race, gender and other aspects of identity. DNA has become a prominent motif in current culture, and is increasingly used in everyday language, often in phrases of the form 'characteristic X is in the DNA of this person/concept/organisation' implying that whatever characteristic X might be, it is essential for that person/concept/organisation. This form of words picks up on the idea that DNA is central to identity. However, any clear definition of identity is elusive, and attempting to locate the essence of humanity in the information carried in DNA is as problematic as using any other defining characteristic (this will be discussed in more detail in Chap. 6). More broadly, since all complex life on earth uses DNA to transmit information between generations, and given that we share significant parts of our genome not just with other animals and plants but also with bacteria and other 'primitive' single-celled organisms, any ideas of human essence and human exceptionalism begin to look blurred. This goes against the general tendency to split humans away from other organisms, and into different groups within humanity, based on what are thought to be natural lines—to 'carve nature at the joints'. The digital nature of the DNA sequence can feed into this, and genetics can be used as a spurious rationale to categorise people, most obviously in terms of ethnicities. Categories are useful in medicine, where there is a need to define conditions to allow treatment, but in reality most biological features fall on a continuum rather than on either side of a clear dividing line. Almost any human characteristic of any complexity is better characterised as being at some point on a continuous spectrum, rather than being assigned to a separate compartment.

One of the striking outcomes of the increase in genomic knowledge has been the unambiguous demonstration of not only our close genetic relationships with other organisms, but also how little genetic difference there is between geographically separated groups of humanity (see Chap. 7). There is no genetic basis for race as that term is commonly used; race is pretty much entirely a historical and social construct with no real biological justification. Differences in the sequence of DNA from a patient compared to the 'normal' gene sequence can be used to give the diagnosis of a genetic disorder (where 'normal' is shorthand for a consensus human genome sequence).[2] However, normal is perhaps not an entirely helpful term in this context. All humans other than identical twins have different genomes, and what is defined as normal will inevitably have a historical basis (i.e. which genomes happened to be in the original sequencing studies). An important bias is introduced when most of the known sequences are from one particular group (the majority of the human genomes sequenced in early studies were of European origin). This can cause errors in assigning disease-causing status to apparently rare variants that turn out to be relatively common in other populations (see Chap. 7). Although humanity as a whole is very homogeneous genetically, all of us carry variants (usually more or less neutral in effect) that differ from the consensus genome at millions of positions, and it is often fairly arbitrary which version is defined as normal (different variants are present at different frequencies in different populations).

Categorisation as female or male is even more pervasive, with an apparently solid biological basis. Genetics perhaps does have something to say about gender, but again great care is needed about making any definitive statements, since although there is a basic biological mechanism behind the division into female and male, the situation can be complex and fluid even at the biological level, without considering the enormous role of society in assigning gender stereotypes (see Chap. 4 for further discussion).

---

[2] The genome is the sum total of all of the genetic material of an organism.

Future Promises and Concerns

DNA sequencing is generally regarded as the gold standard approach for allowing genetic changes to be detected and characterised.[3] Until relatively recently there were significant bottlenecks due to intrinsic limitations on generating sequence data. Essentially, only one fragment of DNA could be sequenced at a time (using Sanger sequencing, see Chap. 7 for more details), which put an upper limit on how much sequence data could be routinely produced in diagnostic laboratories. This upper limit was exceeded by many orders of magnitude by the advent of next-generation sequencing (NGS) over the last decade or so.[4] NGS collectively describes a number of methods that employ massively parallel sequencing—in other words simultaneously sequencing immense numbers of short DNA fragments—coupled with sophisticated software packages that allow assembly of the sequenced fragments into contiguous sequences, right up to whole genomes. The original Human Genome project announced success in 2000, and took years of multinational cooperation (and competition) between research groups and billions of pounds to get to that stage. In a short space of time, we have moved on to the introduction of instruments that can sequence something like 100 genomes in a week, at a cost per genome of a few hundred pounds. This has hugely changed the possible reach of diagnostic genomics, although as will be discussed in Chap. 7, perhaps not by quite as much as advertised. The geneticist Steve Jones has been quoted as saying something on the lines of 'DNA is a molecule made up of four letters; H, Y, P and E'. There is certainly a history of advances in genetics being hailed as revolutionary in medical terms, and then failing to change medicine in anything like the ways promised. This is a real risk in the case of genome-scale sequencing, notwithstanding the landmark scientific success of the original Human Genome Project. Many of the claims made for the benefits of widespread genome sequencing are unsupported by any systematic review of the results to date (of which more in Chap. 7—as David Hume put it, we are wise to proportion our beliefs to the evidence).

Some of the more hubristic claims made for advances in our knowledge of genes and for our technical capabilities might reasonably worry those who

---

[3] DNA sequencing is the term employed for any laboratory process that can be used to establish the exact order of the chemicals that spell out the genetic code in any individual. This allows detection of potential harmful changes in genes in a patient who may have a genetic disorder. The main methods in use today are briefly discussed in this chapter and in Chap. 7.

[4] Next-generation sequencing is not a hugely helpful label; as methods develop further it will become increasingly arbitrary to lump all of the techniques together under one heading. See Chap. 7 for further discussion.

don't look on science (or perhaps how science is applied) as an unalloyed good. There are understandable concerns that the improvements in technology (particularly in DNA sequencing and DNA editing) are moving faster than the ethical discussions around how (if at all) these technologies should be applied. Technical advances have also allowed the relatively new field of direct-to-consumer (DTC) DNA testing to rapidly increase in scope over recent years.[5]

Genetics in Other Areas of Medicine

To avoid giving a false impression, it is worth mentioning that the types of genetic testing covered by this book by no means exhaust the role of genetic testing in medicine. To take just one example, genetic analysis is hugely important in microbiology and virology in detecting bacteria and viruses with great sensitivity and specificity; it is likely that most people first heard of the polymerase chain reaction (or PCR; see below) in the context of COVID-19 testing. Genomic analysis (much quicker for a virus with a genome 100,000 times smaller than the human genome) has also been invaluable in monitoring the evolution of new variants during the pandemic.

Basic Concepts in Genetics

This book is not meant to be a genetics textbook, but an attempt to draw out some of the more important aspects of genomics and genetic testing. The technical basis of the tests won't be explored in full detail—there are many excellent online resources in the field for those interested in the technicalities (see the preface for some examples). However, there are a few terms that probably need to be defined, and some background to be filled in, if the rest of the book is to make sense. Some more specialised terms will be explained where needed for understanding a specific test, but a few basics are covered below, in the hope of conveying the most important concepts.

---

[5] Direct-to-consumer (DTC) DNA testing describes any genetic test that can be accessed by anyone prepared to pay for it, without a medical intermediary. Many of these tests purport to give details on ancestry, and are mostly harmless as long as you don't put too much weight on the results. Tests that offer to establish immediate family relationships such as paternity or to detect clinically relevant changes in the DNA, shorn of any context given by genetic counselling, are more problematic.

DNA Stores Information that Can Be Copied

DNA is the genetic material found in every cellular living organism. Life on this planet, in all of its endless forms going back to the last universal common ancestor, has evolved by using DNA as essentially a massively powerful information carrier. Some of the basic details are covered below, but in terms of genetic disorders and cancers the salient point is that the existence of a more or less precise message implies the possibility of errors affecting that message, and the likelihood that any random or semi-random change will not improve the utility of that message. Such errors are generally termed mutations.[6] The DNA we all share is sometimes described as a blueprint for a human being, but a better comparison is with a recipe, a set of instructions to be followed. The information in this recipe is held as a digital code, made up of the four different chemicals known as bases, adenine, cytosine, guanine and thymine, usually shortened to A, C, G and T. Although your genome, the sum total of your genetic material, amounts to two copies of some three thousand million bases (3×10^9), this is far too few to allow the one-to-one mapping of every cell in the body implied by a blueprint. At each position of the 3×10^9 in the human genome there are four possible alternatives, generating an enormous number of theoretically possible genomes.[7] Although functional constraints limit the number of actually possible genomes, the combinations are enough to ensure that all humans, with the exception of identical twins, start off genetically unique (and even identical twins will accumulate differences at the DNA level throughout their lives).

The genetic material has certain critical properties, which allow both the flow and copying of information. The iconic DNA double helix takes this shape because the four bases can pair with each other, but under normal conditions can only form two pairs: A with T and C with G. This has the consequence that once you have one strand, the sequence of that strand determines the sequence of its partner—the two strands of DNA are described as complementary. This allows DNA to make new copies of itself (for example when cells divide) and most of the time to produce a perfect copy of each gene

---

[6] There is a risk of ambiguity in using this term; sometimes mutation is used to denote any change in comparison to what is accepted as the normal sequence, and sometimes used with the connotation of a damaging change. For the purposes of simplicity, anything referred to subsequently as a mutation will be a change in the genome deemed to be harmful, and the term variant will be used for a change which is benign or neutral, or of uncertain or minor effect.

[7] This capacity of DNA to store information has given rise to serious efforts to explore the use of synthetic DNA molecules as an alternative to electronic information storage (which is limited to only two options, 0 or 1 at each bit). This information could theoretically be stored and reproduced in living organisms, and although the idea is still in its infancy, methods that allow routine editing of DNA may allow this to become a practical possibility.

by re-establishing the base pairs A-T and G-C (although this process is fantastically accurate, errors do occur, and these rare errors can have important consequences).

Genomic Architecture

The human genome is packaged into two sets of 23 chromosomes in each individual, one set inherited from the mother and one from the father. (In addition to the chromosomes located in the cell nucleus, a much smaller but still crucial part of the genome is found in the mitochondria, subcellular structures with a critical role in energy production for the cell; see Box 1.1). Females inherit an X chromosome from each parent, and males a maternal X chromosome and paternal Y chromosome. The non-sex chromosomes (autosomes) come in two sets of 22, numbered (roughly in order of size, chromosome 1 being the largest) from 1 to 22, with one of each pair from each parent (see Fig. 1.1). The numbers are kept constant through the generations by a special type of cell division (meiosis) that occurs in eggs and sperm, ensuring that only one of each pair of autosomes and one X (in an egg) or one X or one Y (in a sperm) are passed on by each parent. The fertilised egg (zygote) will then have the same number of chromosomes as each parent. A chromosome can be thought of as an immensely long strand of DNA, carrying a specific complement of genes and packed into every human cell with a nucleus (see Fig. 1.2). As the chromosomes come in pairs, so do the genes they carry, and each gene of the matching pair is called an allele. The two alleles may be identical, or may differ at one or more positions in their sequence. The allele that represents what is defined as normal for a given sequence is called the wild type allele; this usage reflects the origin of genetics in the study of animal and plant species.

Cells can vary hugely from one tissue to another, but all share the same genome. The differences in cells are due to differences in which genes are switched on—expressed—and which genes are switched off. In addition to there being a characteristic suite of genes active in any particular cell type, within each cell genes are switched on and off in response to environmental signals.

Box 1.1 Mitochondrial Genomes

Not all genetic disorders are due to changes in nuclear genes. Mitochondrial inheritance is involved in a range of severe genetic conditions, and there

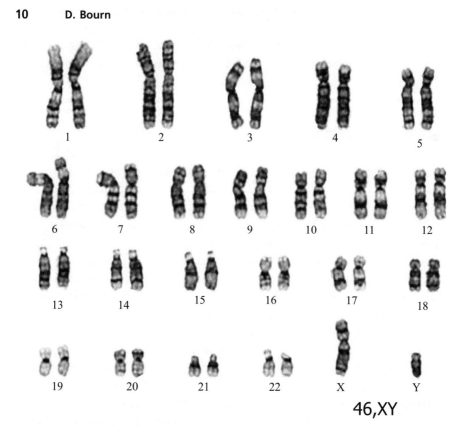

Fig. 1.1 A normal male chromosome complement (described as 46, XY for the total number of chromosomes and the possession of both an X and a Y chromosome). Photograph courtesy of Gavin Cuthbert, Northern Genetics Service

are important differences in the configuration of genes in the mitochondrial genome and in how they are transmitted, as compared to the nuclear genome. Mitochondria are passed on in the egg but not the sperm, and mitochondrial inheritance is therefore exclusively maternal. (This means that the female lineage for any individual can be traced by looking at variants in the mitochondrial DNA—a hugely useful tool in genealogy and in the study of historical human populations.) The mitochondrial genome comprises a single circular molecule of fewer than 17,000 bases, orders of magnitude smaller than the nuclear genome. Of the fewer than 40 mitochondrial genes, most produce a final RNA rather than a protein product. Those proteins which are produced, like the RNAs, have roles in the energy production process in the mitochondria. The organisation of genes in the mitochondrial genome, and to some extent the genetic code, differ from the nuclear genome in ways that reflect the bacterial origin of mitochondria. Although the mitochondrial genes are crucial, and damage to these can result in severe disorders, there are more nuclear genes than mitochondrial genes contributing to the necessary biochemistry in these sub-cellular structures. There are usually multiple

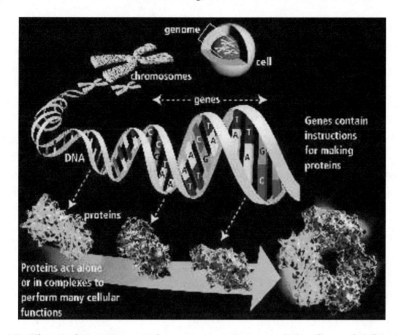

Fig. 1.2 The nuclear genome: chromosomes, genes and the flow of information from DNA to proteins. (Reprinted from U.S. Department of Energy Genomic Science program, https://genomicscience.energy.gov)

mitochondria in a cell, and multiple genomes in each mitochondrion, so in contrast to the single nuclear genome in each cell, there may be thousands of mitochondrial genomes. Mitochondrial genomes harbouring a mutation often coexist with normal genomes within a single cell (this is termed heteroplasmy). The proportions of mutated genomes may differ in different cells, complicating the clinical picture as the threshold for a clinical effect might be exceeded in some tissues but not in others. When mitochondrial genomes are passed on in an egg from a mother who is heteroplasmic for a mutation, only a small subset of the total are inherited (a bottleneck effect). This means that predictions on how severely any children may be affected are difficult, as by chance it could be that mainly normal mitochondrial genomes are inherited, reducing the likelihood of clinical problems, or conversely mainly genomes with a mutation can be passed on, in which case the children may be more severely affected than their mother. Testing for mitochondrial disorders is a highly specialised area, involving a range of disciplines.

Gene Function and Organisation Within Genomes

Although there is no all-encompassing definition of a gene (exceptions exist which defeat any simple categorisation) a gene can be thought of as a stretch of DNA with a specific function, such as coding for a protein.[8] When that function is damaged in some way, a genetic disorder can be caused, as described in subsequent chapters. Protein production is not the only purpose of genes; in many cases a single-stranded RNA is the functional product. Different RNAs have roles in a range of processes, including regulation of gene expression.

Each human gene has an official, universally recognised symbol. For example, the genes implicated in Huntington disease and cystic fibrosis (Chaps. 2 and 3) are denoted *HTT* and *CFTR*, respectively (the convention is that gene names are italicised, and the corresponding protein, such as CFTR, not italicised). In the early days of genetics the assumption was that one gene produced one protein. It is now known that flexibility is the norm, and there are many examples of genes producing multiple different versions of proteins, through variations in the transcriptional processes described below.

Prior to the first sequencing of the human genome, estimates for the total number of genes were often set above 100,000, based in part on an estimate of the number of different gene products in each cell. The correct number turned out to be a little over 20,000. This figure is very similar to that of many other mammals, and not many more than some animals we might think of as more primitive (and plants may have significantly larger genomes than humans, but bear in mind that larger does not necessarily imply more complex). With the benefit of hindsight, this need not have been so surprising. Evolution can only work with what is to hand, and it is more economical and elegant to produce complexity by using existing genetic material in different ways than solely by a brute increase in gene number.

The basic biology of DNA is the same across all of life as currently known, and complexity (although of course sheer size of genomes plays an important part) stems from the way that genes are expressed, not just in terms of amount of gene product but in different temporal and spatial patterns of expression. Gene activity is modulated by environmental signals and by epigenetic factors (modifications controlling gene expression which are not themselves part of the primary DNA sequence, epigenetics in this sense meaning above or on

---

[8] Proteins are the highly complex strings of different amino acids that underpin the bulk of the structures found in living organisms, as well as the enzymes that allow the complex chemistry of life: the properties of a protein are determined by the order of the amino acids that make it up.

top of genetics). Epigenetics will be discussed in more detail in Chaps. 4 and 7.

Our genetic similarity to other forms of life on earth should make humanity a little more cautious about claiming human exceptionalism—it is difficult to draw a clear and consistent dividing line based on genetics that separates all humans from all other organisms. The genes responsible for a number of human genetic disorders were identified in part by studying animal models (other species having similar clinical symptoms, due to a mutation in the equivalent gene to that causing a human genetic disease). This again is suggestive of some close relationships, although of course the differences between species limit the use of this type of analysis.

Another perhaps surprising aspect of the genome is how little is taken up by genes and known gene-associated sequences—of the order of 5%, although it is difficult to give an exact figure (see Box 1.2). Large portions of the remainder of the genome are made up of repetitive elements, with a range of different evolutionary histories and in many cases no apparent function (although comparative genomic studies suggest that most of the genome is functional, we just don't yet understand in what way). Some diagnostic applications for simple repeat sequences are discussed in Chap. 6.

Box 1.2 Gene Regulation, Introns and Exons

It is not just the case that gene-related sequences account for only a small proportion of the genome. For most human genes that produce a protein product, the part of the gene which codes directly for the amino acid sequence of the protein may be only a small part of the whole stretch of DNA needed for expression of that gene. Genes include flanking regions that control the extent and timing of gene expression in response to genetic and environmental messages, there are sequences in the DNA that mark the beginning and end of genes, and sequences that demarcate the split of genes into exons and introns. Exons and introns are important features of human genes (being found in most nuclear genes, but not in mitochondrial genes). Exons contain the coded information for production of proteins, and are interspersed with introns, which can be very much larger than the exons they separate. The DNA making up the introns may contain signals important for the expression of genes, but because this DNA does not carry information which maps directly onto the protein sequence, the intronic material is removed from the products of gene expression. The exons are spliced together to make a continuous sequence containing the information needed to specify the amino acids in the corresponding protein in the process of transcription (see the next section). It follows that a large proportion of mutations leading to human genetic disease are in the exons, which has implications for the approaches used in screening for these mutations—this will be revisited in Chap. 7. Mutations which affect the sequences at the exon–intron boundaries are also important causes of genetic disease. These sequences form part of specific signals that tell the transcriptional machinery where to splice together the exons.

Transcription, Translation and the Genetic Code

Those genes that produce proteins do so via an intermediary, the similar molecule RNA, which also has four bases (A, C and G as in DNA plus U (Uracil) in place of T; as previously mentioned, RNA itself may be the functional product of a gene). Crucially the pairing rules are the same as for DNA, so in the process of transcription (an enormously complex process, greatly simplified here) the DNA double helix is unzipped and used as a template to produce a complementary RNA copy (messenger RNA, or mRNA in protein-producing genes). As part of this process the introns are removed in the formation of the mature mRNA, a step called splicing which requires its own precise signals in the DNA sequence. For protein-producing genes, transcription is followed by translation, where the genetic code now represented in the mRNA is translated (again skipping over the many complexities of the process) into a string of amino acids making up a specific protein with a specific function. The way in which the genetic code conveys information is often compared to the transmission of a meaningful message via a written language. The coding part of the gene is made up of successive codons, which are three bases in length, and codons can be regarded as analogous to words in a meaningful sentence. Given that there are four bases, there are 64 possible codons ($4 \times 4 \times 4$). Most codons are read via the processes of transcription and translation as a specific amino acid; for example the codon ATG is recognised as specifying the amino acid methionine. (ATG is also known as the start codon, as this sequence marks the beginning of the coding region.) A set of 20 different amino acids is needed to produce the proteins that make up most of our body structures, so 64 possible codons give room to specify all of these (most amino acids are specified by more than one codon). This total also allows capacity for three different stop codons (see below) which signal the end of the coding part of the gene just as the ATG codon signals the start of the coding region. A language using two-letter words would have insufficient capacity ($4 \times 4 = 16$) and more than three would be wasteful (and would have been ruthlessly culled by natural selection). In this way, the linear sequence in DNA can produce a precisely corresponding linear sequence of amino acids in a protein. The implication of this is that an error in the DNA sequence, whether acquired or inherited, can cause an alteration in the protein, or even prevent any protein being produced at all. This brings us back to the concept of mutation.

Mutation

As mentioned above, this is another term that defies a simple, all-encompassing definition, but for our purposes a mutation can be understood as a change in the DNA sequence of a gene, either inherited or acquired, that has a deleterious effect on the function of the gene. The term would also encompass loss or gain of genetic material (deletions or duplications respectively) which often can be damaging. The complex flow of information from gene to functional protein or RNA has many stages, and any error in the DNA sequence has the potential to damage the process at any one of these steps. In practice, most genes have variants in their sequences as compared to the consensus sequence, and it can be very difficult to know whether these variants are of any significance. Whether or not a given variant has an effect may depend on the particular genetic background of the individual with that variant, as well as on other environmental factors. This will be a theme covered in more detail later, particularly in Chap. 7. Diagnostic human genetics is essentially a collection of methods for determining whether or not a patient has a genetic change that is causing their clinical symptoms. The scale of the change could range from the loss or gain of whole chromosomes or large regions of chromosomes (almost always with devastating effects as multiple genes will be involved) down to mutations affecting a single base in the DNA (with more varied effects, but in some cases still severe). Some genetic diseases are always caused by the same mutation (the paradigmatic example being Huntington disease, as described in Chap. 2) and some can be caused by a variety of different changes. Different mutations in the same gene can cause different disorders (see Chap. 4 for some examples), and some disorders can be caused by mutations in different genes in different families (termed genetic heterogeneity). A missense mutation involves a sequence alteration that changes the amino acid coded at that position. A single base change in the codon CGA (arginine) to GGA would introduce a glycine amino acid at that position. Whether or not this is deleterious might depend on how different the two amino acids are in their chemical structure, and whether the change occurs at a functionally important part of the protein. Some single base changes don't change the amino acid (recall that most amino acids can be specified by more than one codon). Such changes are usually neutral in their effect, but this is not always the case; in some positions the nature of the base is important not just for the amino acid specified but also for how the messenger RNA is processed. Some single base changes introduce one of the three codons (stop codons, TAA, TAG or TGA) that do not specify any amino acid but signal the end of the coding region of a protein-producing gene. These nonsense mutations cause premature termination of the string of amino acids and in most cases effectively abolish the gene function. If bases are deleted (removed from the sequence) or inserted, the effect may depend on whether or not the change involves a multiple of three bases. Where a multiple of three bases is involved, the mutation will either remove amino acids (one amino acid lost for each three bases deleted) or introduce new amino acids (one amino acid for each additional three bases). The extent to which this is detrimental will depend on the nature of the protein and amino acids involved. However, if a deletion or insertion is not of a multiple of three, this is almost always damaging to the gene function. These are termed frameshift mutations. The concept of a reading frame refers to the way the genetic code stores information. The codons must be read as discrete groups of three bases. The message only makes sense when the codons are spaced sequentially like words in a sentence. Loss or gain of a single base (or any other change not involving a multiple of three) scrambles the message in such a way that the amino acid sequence after the mutation is changed. For example, the short sequence ATG CCG CAG GTA GGG… codes for the amino acids methionine, proline, glutamine, valine and glycine. A deletion of the first base shifts the reading frame by one base giving TGC CGC AGG **TAG** coding for cysteine, arginine, arginine again (a different codon) and then bringing a stop codon (underlined and in bold) into frame, signalling the end of a coding sequence.

In practice, unless the mutation occurs towards the end of the coding sequence of the gene, a change in the reading frame will almost always introduce one of the stop codons downstream of the change, abolishing gene function.

The examples above are not exhaustive; there are many possible ways in which a mutation can affect the way a gene works, some of which will be introduced in future chapters.[9]

---

[9] In addition to the type of mutation, the location of a cell where a mutation occurs is of great importance. If a change is present or occurs as a new event (a *de novo* mutation is the technical term) in an egg or sperm, this is a germline mutation that will be inherited by any child conceived with that egg or sperm, and will be present in all of the cells in that child. In contrast to the germline cells (eggs, sperm and their progenitors), all the remaining cells in the body are described as somatic. A mutation in a somatic cell will only be passed on to the descendants of that cell when it divides (see Chap. 5 on cancer for further discussion). Somatic mutations are not passed on to the

Patterns of Inheritance: Autosomal Dominant and Autosomal Recessive

How a genetic disorder is passed on through a family depends on the nature of the gene function and on whether the gene is inherited on an autosome or the X chromosome (there are very few genes on the Y chromosome, which is principally of significance in sex determination). An autosomal dominant genetic condition is one caused when mutation in a single gene of a pair of autosomal genes is sufficient for symptoms to develop (i.e. an affected parent has a 1 in 2 risk of passing on the mutation and the disorder to any children). An autosomal recessive genetic condition is one caused when mutation of both genes of a pair of autosomal genes is necessary for symptoms to develop (i.e. both parents must be carriers for a mutation and both must pass a mutation on for a child to be affected).[10] The parents would typically be unaffected, and the chance of any child being affected is 1 in 4. The differences will be explored using specific examples in the relevant chapters (Chaps. 2 and 3), but essentially there are some genes where one copy is not enough when the other carries a mutation and some genes (the majority) where one copy is perfectly sufficient for normal functioning. A crucial point that genetic counsellors have to make clear to patients is that the risks quoted above of a genetic disorder being passed on (1 in 2 for an autosomal dominant disorder, 1 in 4 for a recessive) apply to each pregnancy. Each conception is an independent event, and the risk for any child is not affected by what has happened before. The definitions given above only apply where mutation(s) in a single gene are sufficient to cause a genetic disorder: patterns of inheritance do not always correspond to these basic models.

X-linked Inheritance

For a female with two X chromosomes the situation for genes on the X chromosome, could be regarded as basically similar to that described for autosomes. However in males, who only have one X chromosome, if a mutation is inherited in an X-linked gene (which just means a gene on the X chromosome) there is no compensating copy, and a gene which might act as recessive

---

next generation. It may be of huge consequence to an individual if a somatic mutation occurs in a gene which can predispose to skin cancer, and this event takes place in a skin cell. It would be of no consequence if (for example) a mutation in the gene causing Huntington disease (see Chap. 2) occurred in the same cell.

[10] Carriers of a mutation are described as being heterozygous, having that mutation in only one copy of the gene, as opposed to homozygous, where both alleles have a mutation.

in a female carrier can cause a genetic disorder in a male. Conditions such as Duchenne muscular dystrophy therefore predominantly affect boys. There are additional complexities for X-linked inheritance that are not captured by the brief description above: some of these are discussed in Chap. 4.

Some Basics of Genetic Testing

Although technical methods are constantly evolving, a relatively small number of techniques account for most of the tests currently carried out in diagnostic laboratories. The main historical division in diagnostic genetics was between cytogenetic testing and DNA-based testing. Cytogenetics is very much the longer established discipline, initially involving the examination of chromosome preparations, using the techniques of cell culture and light microscopy. Aspects of chromosome analysis are introduced briefly in Chap. 5 in the context of cancer testing. The division between the disciplines is becoming increasingly blurred as more testing moves to DNA-based methods. The focus of the remainder of this chapter is on methods for manipulating and investigating DNA, aiming to provide a brief introduction to some principles that are essential to understanding the genetic testing covered in subsequent chapters.

Isolation of DNA

In order to perform genetic testing, the first step is usually separating DNA from a patient sample using various techniques for DNA extraction. Any tissue with nucleated cells contains DNA and different tissues may be used for different purposes (for example in cancer testing tumor material will be analysed), but for most inherited conditions accessible sources of cells such as saliva, cheek cells and particularly blood would be the first choice. Most DNA extraction is now automated, and the basic process usually involves as a first step adding chemicals that lyse (burst open) the cells. The DNA is then separated from the cell debris (using centrifugation, or for automated methods sequestered by binding the DNA to a suspension of small magnetic beads). The purified DNA is then dissolved in a small volume of a solution suitable for long-term storage (DNA can be kept in a usable form for many years when frozen at very low temperatures, typically between -20 and $-80\ ^{\circ}C$).

Finding Mutations

Various approaches can be used to detect mutations in a single gene against the background of the whole genome, and specific examples will be mentioned in the context of testing for individual genetic disorders. Very broadly, targeted testing methods can be split into those that detect the region of interest using variations on the theme of hybridization, taking advantage of the complementary nature of DNA, and those that detect the region of interest by making a very large number of copies of that region (amplification). The approach used will depend on the nature of the mutation of interest, the number of samples to be tested and the resources available.

Finding a Complementary Sequence

Assays using hybridization rely on the complementary nature of DNA. In the most basic approach, DNA from a patient including the gene or genes of interest is made single-stranded. A stretch of single-stranded DNA with a complementary matching sequence to the region of interest is made in the laboratory and labelled to allow detection (usually radioactively when this process was originally developed, usually now using fluorescent chemicals to tag the DNA). This is termed a probe. This can then be used to find the complementary sequence in the extracted patient DNA, forming a hybrid molecule between the probe and the target patient DNA. Although useful as an early tool in DNA analysis, this method as first applied was relatively limited in scope, and as sample numbers increased became too slow and cumbersome for efficient diagnostic work, apart from a few specialist applications.[11] The principal techniques that now employ hybridization allow an improved throughput and much better resolution. Typically a huge number of short, single-stranded DNA molecules are attached to a solid support (an array or microarray). Each of these is complementary to a different region of the human genome, and the array of molecules is designed as far as possible to interrogate regions across the whole genome (with additional markers for regions known to be important in human genetic disorders). Suitably treated patient DNA (fluorescently labelled and made single-stranded) can then be

---

[11] Early DNA-based testing often used the process known as Southern blotting, named for its inventor Sir Edwin Southern. This technique relied on fractionating human genomic DNA in a highly specific way before separating the fragments by size using electrophoresis (see below) and transferring these to a membrane before hybridizing with a radioactively labelled probe for the gene of interest. Although still used for specific applications where it is necessary to detect very large DNA fragments, the technique has largely been superseded by PCR and other methods.

hybridized to these many probes. The pattern of hybridization can be analysed to show gains or losses of chromosomal material at a much better level of resolution than light microscopy.

The Polymerase Chain Reaction and DNA Amplification

It is difficult to overstate how much difference the development of the polymerase chain reaction (PCR) in the mid-1980s made to diagnostic genetics. It still underpins many of the tests in use today. I won't go into the full technical details (there are many excellent online tutorials for this and other aspects of genetic testing) but essentially the method allows laboratories to make a huge number of copies of a short region of DNA from a patient (often a few hundred bases of DNA for diagnostic work) which includes the position(s) of interest. After repeated rounds of amplification the specific product is present to such a high level that it can be analysed without any interference from the rest of the patient's DNA. The DNA copying is rapidly performed in the laboratory using one of a class of enzymes called DNA polymerases, which synthesise new DNA strands by incorporating the four bases in the correct order, using an existing DNA strand as a template. This amplified region is labelled by the addition of chemical groups that allow easy detection (usually fluorescent tags in current methodology). This facilitates analysis of gains or losses of small lengths of genetic material, can be used to directly analyse single base changes, and also lends itself to use in Sanger sequencing, for the determination of the exact sequence of DNA bases in the amplified fragment of DNA (see below).

Electrophoresis

Later chapters will include several examples of genetic tests where it is necessary to determine the size of a fragment of DNA from a patient (a typical example is in the sizing of DNA fragments initially generated by PCR), and the general process of sorting biomolecules by size and by charge is termed electrophoresis. In the case of DNA in solution, the molecules have a uniform negative charge along their length as a consequence of their chemical structure. If placed in a suitable matrix and exposed to an electric current between two electrodes, fragments of DNA will migrate towards the positive electrode. Smaller fragments will be physically less impeded by the matrix, and will

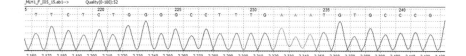

Fig. 1.3 An example of Sanger sequencing. PCR-generated fragments are subject to additional amplification using a technique that labels each of the bases in the fragment with a fluorescent tag (the four bases in DNA are labelled in four different colours) allowing detection after capillary electrophoresis. See Chap. 7 for further details

move faster. It is therefore possible to separate DNA fragments by size, and this can be done accurately enough to distinguish a difference of a single base in fragments several hundred bases long. Initially, this was accomplished using gels something like slabs of jelly made in a mould (library footage from genetics laboratories inevitably includes a laboratory technician loading coloured liquids into a gel for the benefit of the cameras). An example of this type of electrophoresis is shown in Fig. 4.2. More accurate separation was achieved using very thin gels formed between two glass plates, but these were physically difficult to produce and fiddly for loading samples. Electrophoresis is now characteristically done using large arrays of fine capillary tubes and capillary electrophoresis has a number of advantages in terms of speed (being amenable for inclusion in automated processes) and accuracy (see Fig. 1.3 for the first of several examples of the output from capillary electrophoresis). As shown in this figure, the technique also allows for simultaneous detection of multiple fluorescent dyes, ideal for Sanger sequencing.

DNA Sequencing

It is now taken for granted that sequencing DNA from any organism can be done routinely, cheaply and increasingly at tremendous volume. It has taken many increments in the technology to get to the point where, as mentioned above, multiple human genomes can be sequenced in days on a single instrument. The sequencing done in diagnostic labs can be broadly divided into two general approaches: a low throughput piecemeal workflow using Sanger sequencing, based on sequencing a PCR-generated fragment or fragments (see Fig. 1.3), and the collection of techniques under the heading of next-generation sequencing (NGS). Sanger sequencing and NGS will be described in more detail in Chap. 7.

Hopefully, this chapter has provided sufficient background to the basic science and basic technology underlying diagnostic genetics. Some additional details are given in the following chapters in the context of specific tests.

2

Autosomal Dominant Inheritance and Huntington Disease

Summary This chapter explores aspects of genetic testing in the context of autosomal dominant disorders. Huntington disease is used as an example of an autosomal dominant disorder caused by a trinucleotide repeat expansion, which gives rise to a toxic gain of function. The dynamics of repeat expansions are touched on, and the relevant laboratory testing methods are described. A second condition, familial hypercholesterolaemia caused by *LDLR* gene mutations, is used as an example of an autosomal dominant disorder with a different and perhaps more representative loss of function mechanism. Clinical scenarios leading to investigation for Huntington disease are addressed together with the utility of the tests. Some more general aspects of laboratory testing are also addressed, including sensitivity, specificity and sources of laboratory error. Ethical questions raised by testing for this late-onset disorder are highlighted at the end of the chapter, largely around the sharing of genetic information.

Huntington Disease

Huntington disease (HD) is in some ways one of the most clear-cut (as well as one of the most devastating) single gene disorders, in that genetic determinism seems to apply—if you inherit a mutation in the relevant gene you will inevitably develop the condition. As we shall see, even in this case there are some grey areas (a frequent theme of genetic testing, even for well-understood disorders, is the gradual discovery of levels of complexity). HD is relatively common as genetic conditions go, estimated to occur in up to

© The Author(s), under exclusive license to Springer Nature Switzerland AG 2022
D. Bourn, *Diagnostic Genetic Testing*,
https://doi.org/10.1007/978-3-030-85510-9_2

one individual in 10,000 depending on which population you look at. It is an autosomal dominant disorder, which means that inheriting one damaged copy of the gene is sufficient to cause clinical symptoms (see Chap. 3 for a contrast between dominant and recessive disorders).[1] HD is always caused by the same genetic error. The protein (Huntingtin) produced by the gene in question (the *HTT* gene on chromosome 4) plays an important role in nerve cells, and the clinical features of the disorder reflect this, including problems with movement, cognitive impairment and psychiatric disturbances. The damage is progressive and currently irreversible.

A Very Specific Genetic Error

The mutation that causes HD is a genetic error of a type known collectively as trinucleotide repeat expansion disorders (a trinucleotide can be thought of as equivalent to a run of three bases, although the terms nucleotide and base are not strictly interchangeable).[2] Some genes contain regions where a sequence of three bases repeats itself many times. In the case of the *HTT* gene, the repeated sequence is CAG. In the genetic code CAG corresponds to the amino acid glutamine, so this run of CAG repeats produces a run of glutamines in the resulting protein (this type of disorder is therefore also known as a polyglutamine disorder). In a functioning protein produced by the *HTT* gene there can be between 6 and 35 consecutive CAG repeats. However, if more than 35 CAG repeats (and therefore more than 35 glutamines) are present, HD can result (hence trinucleotide repeat expansion disorder: the condition is caused by an increase in length of the repeat array).[3] The reason why this apparently small difference above a threshold can have such a major effect isn't obvious, despite a great deal of research in this area. The only mutation which can give rise to HD is the expansion in the CAG repeat found in the *HTT* gene.

---

[1] Conditions like HD are described as having age-dependent penetrance: you are more likely to develop symptoms the older you get. Penetrance is a general measure of how likely you are to develop symptoms given that you have inherited a particular mutation. This can be modified by, for example, other genetic factors or environmental triggers and will be different for different disorders. A mutation with 50% penetrance would be expected to cause clinical symptoms in half of the individuals who inherit that mutation. Most of the mutations seen in HD have approaching 100% penetrance.

[2] A nucleotide is made up of a base, A, C, G or T in the case of DNA, which is the information-carrying part of the molecule, plus a 5-carbon sugar, deoxyribose, which together with phosphate groups makes up the backbone of the DNA double helix.

[3] This represents a slight oversimplification as the total run of glutamines in the protein includes that produced by a CAA codon, which also codes for glutamine, and is followed by another CAG at the end of the run of repeats. For clarity's sake I will only refer to the uninterrupted run of CAG codons when talking about repeat numbers.

A Gain of Function

There are innumerable ways in which a mutation in the DNA can damage a gene. You might therefore expect to find many different mutations among patients with a particular genetic disorder, the only thing in common between the mutations being disruption of part of the process of gene expression, or of some aspect of the function of the gene product. For many conditions, it is in fact the case that a range of different mutations (missense, nonsense, deletions, duplications, etc.) are seen. Box 2.1 briefly describes an autosomal dominant condition (familial hypercholesterolaemia due to *LDLR* gene mutations) where this applies. In contrast, where a genetic condition is always associated with a single type of change, this may imply that rather than a loss of function, the change is introducing a gain of function. The protein produced by the *HTT* gene with a run of more than 35 glutamines has gained an additional and damaging function. Despite intensive investigation, it is still not completely clear why this additional function causes HD, but the problems are related to toxic protein deposits in specific areas of the brain. In most autosomal dominant conditions, homozygotes (unlucky enough to have inherited two copies of a damaged gene, where even one damaged copy is sufficient to cause clinical symptoms) are more severely affected. This is less obvious where a gain of function is involved; it is unclear to what degree inheriting two expansions in the HD gene worsens the condition.

Box 2.1 Familial Hypercholesterolaemia

HD is always caused by the same genetic error. In contrast, there are multiple mutations in the low-density lipoprotein receptor (*LDLR*) gene on chromosome 19 that can cause the autosomal dominant condition familial hypercholesterolaemia (FH). The normal function of the LDLR protein is to mediate cholesterol uptake.[4] The many known mutations (well over a thousand have been reported) cause a loss of function in various ways. FH is a dominant condition because one functional copy of the *LDLR* gene is not sufficient to prevent harmful accumulation of cholesterol (there are other genes associated with this phenotype, which also show a dominant mode of inheritance). The resulting high cholesterol levels lead to an increased risk of cardiovascular disease, and FH is common enough to make the overall health burden due to this condition significant (cumulatively mutations in *LDLR* and other genes causing FH are present in as many as 1 in 200 of the population).

---

[4] The discussion above is somewhat simplified as the low-density lipoprotein receptor is only involved in uptake of one of the main types of cholesterol, and although high cholesterol levels are a good proxy for cardiovascular disease risk, the situation is more complex than a straightforward causation.

Children of an affected individual have a 1 in 2 chance of inheriting the mutation and therefore being at risk of symptoms. As FH due to *LDLR* mutations is sufficiently common, homozygosity (where individuals have inherited a mutation from both parents) is also seen more frequently than is usually the case for a dominant disorder, and these individuals experience severe and early coronary artery disease, often dying in their second or third decade of life. Given that FH is (in heterozygotes) a condition of later life, and is more problematic where access to a high fat diet is easy, this is an example of a disorder where the clinical problems become much more apparent in a particular environment. In circumstances where life expectancy was shorter and other risks higher, having lower *LDLR* expression could have been benign or even advantageous (for example high cholesterol has been suggested to have a protective effect against infectious disease, a major mortality risk for much of human history).

Initially, testing for *LDLR* mutations (and mutations in other genes associated with FH) tended to be targeted assays which detected some of the commoner changes, which had the advantages of speed and low cost, and only detected changes of known significance. However, this approach will miss some mutations, and increasingly now testing is carried out by sequencing of a panel of different genes known to be involved in FH. In addition to the dominantly acting mutations, there are variants in these genes that individually add a small increment to the risk of high cholesterol, but if present in combination with other variants can cumulatively predispose to a significant increase in cholesterol levels. Full sequencing will also detect these and give the possibility of assigning a polygenic score (see Chap. 7).

FH is a genetic disorder where testing not only informs diagnosis but also can help treatment (unfortunately this is only the case for a minority of genetic conditions). Cholesterol levels can be managed to some degree by adopting a healthy lifestyle and diet, and increasingly drug interventions are possible. Genetic testing helps in the diagnosis for symptomatic individuals, and crucially allows monitoring and potentially early treatment in relatives who can be tested for the familial mutation before they have any symptoms.

Why Expansions?

Simple repeats in DNA are often inherently unstable. In part, this is due to the ease with which errors occur when repeat sequences are copied, as happens during cell division. The process of replication involves unwinding the double helix and using both strands as templates for the production of a new strand. It is relatively easy for the new strand to become misaligned, with resulting loss or gain of a whole number of the repeat units. At the risk of pushing the analogy with language too far, this process could be thought of as similar to errors that might creep in when copying by hand a long, repeated sequence of the same word or phrase. Such errors can lead to the loss or gain of a number of elements of the original text. Trinucleotide repeat expansion disorders occur if the number of repeats increases beyond a threshold when transmitted to the next generation. It has been shown that there can be a bias towards expansions in these repeat regions, and this tendency for expansions may depend on which parent passes on the gene. In the case of HD, large expansions tend to occur on paternal rather than maternal transmission. There are different trinucleotide disorders where the opposite is the case, and yet others where to date no such bias has been found. In general, larger repeats are more unstable.

Determinism, but with Complications

Human characteristics fall on a continuum from entirely genetic to entirely environmental, with the vast majority being shaped by some complex combination of the two. The underlying cause of HD falls close to the entirely genetic end of the spectrum, much more so than is the case for most genetic variants. If you inherit more than 35 CAG repeats you will develop HD—with certain qualifications. Firstly, as mentioned above, penetrance is age-dependent. HD is usually a disorder of later life, with the onset of symptoms typically between 35 and 50 years of age. Secondly, the number of repeats can modify penetrance and affect age of onset. If the number of CAG repeats is between 36 and 39, this is described as being in the zone of reduced penetrance. While repeats of this length are found in affected individuals, they are also found in unaffected individuals in their 70s or even 80s. The factors (environmental or genetic) that determine whether or not 36 repeats will produce symptoms at a given age in some individuals but not others, which are clearly of enormous importance for affected families, are not understood. There is also an inverse correlation between repeat length and age of onset: the higher the repeat number, the earlier (on average) the onset. In particularly devastating cases, individuals with more than 60 repeats can show juvenile onset of the condition. This sort of relationship between the specific version of a gene (the genotype) and the clinical condition or other characteristic (the phenotype) is often referred to as a genotype–phenotype correlation, a recurring theme for genetic disorders.

Anticipation

The tendency for trinucleotide repeat arrays to increase in length when passed on, coupled with the correlation of greater repeat length with earlier onset, gives rise to the phenomenon of anticipation. Clinical symptoms tend to get more severe in successive generations—yet another burden on the families involved.[5] This tendency towards expansion of the repeats in the *HTT* gene provides part of the explanation for the (relatively) common occurrence of the disorder. You might expect a mutation causing severe effects to be lost from the population over time. However, if there is a permanent pool of alleles that can potentially expand (intermediate alleles; see below), the disorder will always be a risk for some individuals. In addition, in the case of late-onset disorders like HD, individuals with the expansion may well have children before they develop symptoms, and therefore possibly pass on an expansion before they even know that they themselves are at risk. For most of human history life expectancy was such that most people wouldn't live long enough to develop symptoms, even if they survived long enough to have children.

Genetic Testing for HD

The nature of the mutation causing HD, and the fact that it is invariably the same mutation, simplifies the task of a laboratory performing diagnostic and predictive testing. Testing is carried out using the polymerase chain reaction mentioned in Chap. 1—a technique that allows the laboratory to produce huge numbers of fluorescently labelled copies of a region of interest, so that a small fragment of DNA can effectively be examined without interference from the rest of the patient genome. The specificity of PCR, ensuring that only the region of interest is amplified, and the rest of the genome ignored, rests with the use of PCR primers, which are short (usually around 20–25 bases), single-stranded fragments of DNA designed by the laboratory for each specific PCR and commercially manufactured. The primers are designed to be complementary to sequences at either end of the region of interest, and to nowhere else in the genome, which allows replication of just the sequence between the two primers. In the case of HD, the primers are designed to recognise sequences just outside the CAG repeat region (see Fig. 2.1). Because

---

[5] In general, care is needed before drawing conclusions from observations that children are more severely affected than their parents. This could be due to ascertainment bias: for obvious reasons it is the case that more severely affected individuals are more likely to come to the attention of the

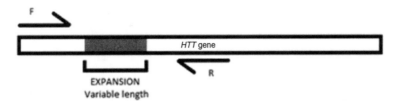

Fig. 2.1 The arrows labelled F (Forward) and R (Reverse) denote the position of the two specific PCR primers (which are complementary to opposite DNA strands). The amplified region demarcated by these primers includes the CAG repeat, shown in red, and the length of the amplified DNA fragment will therefore depend on the number of CAG repeats present

the amplified fragment of DNA includes the repeats, the length of the fragment will differ for alleles with different numbers of repeat units. The process of capillary electrophoresis is used to separate and size the amplified fragments from a patient, and by a comparison with control samples that have alleles of known size (normal and expanded), the size of the CAG repeat on each allele of the *HTT* gene can be determined (see Fig. 2.2).

Test Sensitivity and Specificity

The value of a laboratory investigation is often measured in terms of the test sensitivity (the proportion of real positive cases picked up) and the test specificity (discrimination between positive and negative cases). A test with low sensitivity will fail to detect all positive patients (giving a high false negative rate) and a test with low specificity will give positive results in unaffected patients (a high false positive rate). The ideal test will approach 100% sensitivity (all true cases detected) and 100% specificity (no false positives). In practice, laboratory tests generally don't reach either of these marks, and the uncertainty post-test is increased in proportion to how far short of 100% the test falls for these two measures (much discussed during the COVID pandemic). Genetic tests usually score well for specificity (if you are detecting the presence or absence of a specific mutation with a properly conducted assay, this should come close to 100% specificity). Sensitivity for diagnostic genetic investigations is more variable, depending on the method used and depending even more on the range of different mutations that can affect a particular gene. In the case of HD, with a simple reliable test and the same

---

medical profession. In the case of HD and allied disorders it has been demonstrated that when you control for ascertainment bias, anticipation can still be shown to be real.

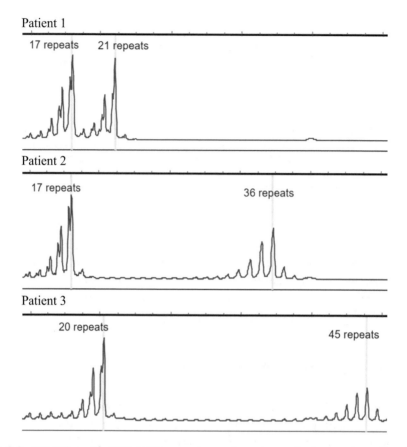

Fig. 2.2 Detection of *HTT* CAG repeat numbers in patient samples after PCR amplification and separation by capillary electrophoresis. Fragments are labelled fluorescently and are detected as blue peaks. Larger fragments (with a higher number of CAG repeats) are found further towards the right hand side of the figure. Each panel shows the result for one patient. Patient 1 has two normal alleles with different repeat numbers (17 and 21 repeats). Patient 2 has alleles with 17 and 36 repeats (36 repeats corresponds to the lowest repeat number unambiguously associated with clinical symptoms) and Patient 3 alleles of 20 and 45 repeats. If Patient 3 were being tested because of a family history of HD, but had no symptoms at the time of testing, this patient would be predicted with almost 100% certainty to develop HD

mutation in all cases, both sensitivity and specificity should approach 100%.[6] However, even given the fact that a test is technically excellent, that doesn't

---

[6] Nothing ever reaches 100%; there is always the possibility of human error, or an extremely rare and unpredictable set of confounding circumstances. In the case of HD, one confounding factor is the possibility of only detecting an allele of one size, usually because the patient is homozygous for normal alleles with the same repeat length, but possibly because a large expansion is being missed. There are second line tests that can be used to resolve this.

mean it should be used. The result also has to be useful in informing and treating the patient. What is the utility in testing for HD, where no good treatment is available?

The Value of Genetic Testing for HD

The most straightforward reason to test for HD is to establish or confirm a diagnosis, which can be done with almost 100% confidence for this disorder. Patients who have symptoms suggestive of HD can therefore be given a test that will reliably tell them either that they definitely do have the condition, if they have an allele with 36 or more repeats, or that HD has been definitively excluded if they have two alleles with repeats falling into the normal range of 6–35 repeats. Even in the absence of treatment, it is important for a patient to know if they do have HD, with the implications for their family that this entails, or whether there is a need to undertake further investigations to establish a different diagnosis.

The test for HD, in common with many other genetic tests, has an additional value in presymptomatic or predictive testing.[7] The number of CAG repeats inherited is determined at conception, so it is possible to carry out a test for HD potentially decades before any clinical symptoms are experienced. Given the frightening implications of a positive result (an incurable, progressive and very distressing condition, with the risk of having passed this on to your children) a rigorous system of genetic counselling has to be in place before testing is carried out. Even a good result can have psychological implications, with feelings of guilt at having been spared when other family members are affected. If you have an affected parent you have a 1 in 2 chance of inheriting an expansion and becoming affected yourself—a coin toss with enormous implications. A majority of those in this situation choose not to have a predictive test, preferring to live with uncertainty. Others would rather know one way or the other, but if you get a bad result you can't then decide not to have the information. Counselling must ensure that the patient is fully aware of the implications, which extend beyond direct health concerns. HD is

---

[7] The two terms are often used more or less interchangeably but do have slightly different meanings. Strictly speaking presymptomatic testing refers to conditions like HD where having inherited the mutation, a patient will develop symptoms, while predictive testing refers to conditions where finding a mutation indicates a predisposition to clinical symptoms without certainty that they will develop. Predictive testing will be described further in the context of inherited cancers in Chap. 5.

unique in that predictive test results may have to be disclosed when taking out life insurance in the UK, but only where the insured sum is over £500,000.[8]

As for other inherited conditions, prenatal diagnosis (PND) is possible for HD, by testing a placental sample (the placenta is tissue of fetal rather than maternal origin) or by sampling fetal cells in the amniotic fluid. However, prenatal diagnosis is relatively unusual for late-onset conditions, and perhaps even more controversial in this context. (Some of the major ethical problems raised by prenatal diagnosis will be discussed in Chap. 3.) As mentioned above, individuals with an affected parent often choose not to find out whether they themselves have inherited an expansion. However, they have to live with the constant worry that they may develop symptoms, and if they have children, there is also the anxiety about having transmitted the mutation to the next generation. If you have a 1 in 2 risk of inheriting the mutation in the *HTT* gene, then any child has a 1 in 4 risk. It is possible in some circumstances to get information about the risk to a child without changing the 1 in 2 risk to the parent (referred to as exclusion testing, see Box 2.2). The language of risk can be confusing—no one is truly at a 1 in 2 risk of HD, you either inherited the mutation or you did not. The use of probabilities just acknowledges a current state of ignorance, although not quite in the same way that the result of a coin toss remains at 50:50 until you know whether it is heads or tails after the event. The difference is that for HD and other genetic conditions the reality, unknown until symptoms develop or a test is carried out, has already been decided at conception.

Box 2.2 Exclusion Testing in HD

Suppose you have a situation where an individual has an affected father, and she herself is pregnant. She must have inherited one of the paternal *HTT* genes, together with a maternal *HTT* gene, and it is 50:50 as to whether she inherited her father's expansion. She does not wish to have a predictive test herself, and based on the current situation any child is at a 1 in 4 risk. However, if the future child had inherited one of their maternal grandmother's *HTT* genes from their mother, rather than one of the *HTT* gene copies from their maternal grandfather, the affected individual in the family, they would be at negligible risk. If they have inherited the *HTT* gene from the maternal grandfather via their mother they would be at the same risk as her (1 in 2). It is possible to do a prenatal exclusion test, tracing the inheritance of the *HTT* gene copies between individuals in a family using variable DNA markers (covered in more detail in Chap. 6). Such tests are rarely requested, and the problem for the

---

[8] Although details of a medical history are of course used by insurers to set premiums, in the UK there is no obligation to disclose the results of any other predictive genetic tests, and the risk is effectively shared across the population.

family comes not when you get a good result—the child from that pregnancy being at negligible risk—but when the news is bad and the child is now at 50:50 risk. Where this is the result, it is not unusual for the prospective parent with the family history (the mother in the example above, but it could equally be the father) to decide to have the definitive test, and thus get a definitive answer for the pregnancy, rather than terminate a pregnancy on the basis of a 1 in 2 risks.

Laboratory Errors

Diagnostic laboratories have a responsibility to provide a safe (and timely) service to the patient; great damage can be done if the wrong result is given. This responsibility applies to any diagnostic testing in any medical laboratory, and there are many potential sources of error, from a mix up with a different patient when taking the original blood sample, through sample mix up or testing error at any stage of what may be a complex laboratory process, to the final interpretation and reporting of the result. In the context of a predictive test for HD, if a sample mix up occurs and a patient is wrongly told that they are not at risk of developing HD, the error is unlikely to be detected until they develop clinical symptoms, possibly years or even decades later. By that time they may unwittingly have passed the mutation on to their children. Errors in predictive testing could have equally disastrous consequences when testing for predispositions to cancer or to life-threatening cardiac defects, and in the special case of prenatal testing could lead to the termination of a healthy fetus or conversely to the birth of a child with a devastating genetic disorder.

If you are having a biochemical or microbiological test you may give a series of samples over time, and there is a chance that a sample mix up will be picked up owing to an anomalous result. For an inherited condition genotyping is usually done only once—the result should be the same irrespective of when the sample is taken and tested—so there may not be the chance to pick up an error until after the consequences have become evident. This also makes it difficult to investigate and eliminate the root causes of errors if they occurred many years previously. DNA-based testing generally involves multiple steps, with the transfer of very small volumes of sample from one stage to the next, so checks are necessary at each stage to minimise errors. Automation is used wherever possible, although this does not completely eliminate the possibility of systemic errors. In practice, there is a great deal of internal and external oversight of laboratory procedures and the error rate is low, but not zero. When an error is discovered it is imperative that it is acknowledged, and in addition to a full apology and explanation, the least that is owed to the patients involved is an undertaking to investigate the mistake and change practices so as to avoid any repetition.

Another imperative for laboratories is dealing appropriately with uncertainty regarding a result. This may be less of an issue in genetic testing where the result is binary (mutation present or absent) compared to tests where there may be a range of possible results, with a potential overlap between normal and abnormal ranges. In the case of HD, by using appropriate control samples (previously tested and checked samples with known repeat sizes) as a comparison it should be possible to give a precise number of repeats, and given the threshold for developing symptoms at 36 repeats it is critical to be able to distinguish 35 from 36 repeats in a patient.[9] It is also important to be able to distinguish so-called intermediate alleles (27–35 repeats, which do not cause HD but may have the potential to expand into the disease-causing range when passed on).

Genetic Information and Families

Suppose you have a grandparent with HD, and you have seen the effect the condition has had on that person and the rest of the family. You yourself are at a 1 in 4 risk, and one of your parents, the child of your affected grandparent, is at a 1 in 2 risk. It might be the case that you wish to be tested, to know for certain, but your at-risk mother or father does not want to know. If you are tested and found to have inherited the mutation present in your grandparent, that mutation has been passed on via your parent, so they now also know they will develop HD.[10] How do you balance your right to have a test with the right of another member of the family not to be given unwanted information? In practice you can't refuse to provide a predictive test to an adult who understands the implications of that test. Genetic counsellors will work closely with the family to try to resolve any contradictions, but there is no easy answer, and there are serious ethical questions regarding who should own genetic information, given that the implications go beyond the individual being tested. This was highlighted in a particularly tragic recent case

---

[9] Although there are some possible reports of symptomatic individuals with fewer than 36 repeats, the consensus is that this threshold is of real significance.

[10] If you have not inherited the mutation, then although your parent remains at risk, their own risk would be reduced by the fact of having had an unaffected child—the next chapter touches on risk calculations in the light of new evidence. Intuitively, it seems right that if an individual who is at an initial 50% risk of HD has a large number of unaffected children, this must reduce the likelihood that the individual in question did inherit the repeat expansion.

involving a family where a man had murdered his wife and was subsequently diagnosed with HD. He refused to share this information with his daughter who was pregnant as he feared she would terminate the pregnancy. His doctors respected this wish and did not inform his daughter. She was diagnosed with HD some years later, and began a legal case against the medical professionals involved, on the grounds that she should have been warned of her own risk to allow her to make a decision regarding her pregnancy. Should health-care professionals have a duty to warn at-risk individuals? This would give rise to immense ethical and practical problems. It is acknowledged that although confidentiality is a central principle of medical ethics, it is not absolute. What circumstances would justify breaking confidence? Where does the avoidance of potential harm outweigh the claim to confidentiality? Professional guidance in the UK, in line with the legal position, is that healthcare professionals are obliged to undertake a balancing act in the context of genetic testing. They must consider the possibility of disclosing confidential information without consent, if a close family member is at risk of serious harm.

3

Autosomal Recessive Inheritance and Cystic Fibrosis

Summary Autosomal recessive inheritance is contrasted with autosomal dominant inheritance. Cystic fibrosis is the main example chosen of an autosomal recessive disorder, and the mutation spectrum is described. The persistence of recessive conditions in a population is explained with reference to heterozygote advantage and founder effects. (These concepts, and that of penetrance, are explored further in a brief section on the recessive disorder hereditary hemochromatosis due to *HFE* gene mutations.) A testing strategy for cystic fibrosis is introduced together with one of the basic testing methods. Some of the multiple referral reasons for cystic fibrosis testing are described, (including neonatal screening and carrier testing) alongside the utility of these investigations. A brief overview of prenatal diagnosis emphasises the deep ethical questions raised by such testing. New specific therapies for cystic fibrosis are mentioned. The final part of the chapter discusses elements of risk and uncertainty, and the use of Bayesian statistics to refine probabilities.

Contrasting Dominant and Recessive Conditions

In the case of some autosomal genes, only having one working copy is not enough. This could be because one copy just does not produce enough functional protein for the relevant purpose, or because the gene product has to combine with others in the correct ratios to work properly (increases in gene copy number, due to a duplication of a gene, can also potentially give rise to a dominant disorder). As we have seen for Huntington disease, a dominant genetic disorder can occur because the mutated allele takes on a new and damaging function. There are also cases where the product of the mutated allele interferes with the product from the normal allele in a detrimental way (a dominant negative effect). In other genes regulation of the normal function is affected such that the gene is expressed in the wrong amounts, at the wrong time or in the wrong tissue—all mechanisms potentially leading to a dominant inheritance pattern. However, for most autosomal genes, loss of function from one allele is not sufficient to give rise to clinical symptoms. Where loss of both functional alleles is necessary to cause a disorder, the condition is described as recessive.

Cystic Fibrosis

Cystic fibrosis (CF) is an example of a recessive condition since having one functional copy of the gene involved (the *CFTR* gene on chromosome 7) is perfectly adequate. Carriers of CF are therefore unaffected; only individuals who inherit a mutation from both parents will develop clinical symptoms. CF is particularly common in Northern Europe, for reasons which will be explored below. The *CFTR* gene product (the CFTR protein) is involved in the transport of ions across cell membranes, and faults in this process lead to tissue secretions becoming too viscous. This causes damage in a range of organ systems, predominantly the lungs, pancreas, gastrointestinal and reproductive systems, and CF is therefore a very serious disorder, although treatment has improved greatly in recent years.

Many Different Genetic Errors: Some with Variable Effects

Although one specific genetic change is by some margin the most common, particularly in European populations, CF is an example of a disorder caused by a loss of function, and many different mutations can be causative. Estimates of the number of different pathogenic (disease-causing) changes in the *CFTR* gene vary, but the total is thought to be more than 1000. CF can equally well occur whether the affected individual has the same mutation in both *CFTR* alleles (i.e. is homozygous for that mutation) or has inherited a different mutation from each unaffected parent (described as being a compound heterozygote). Most of the changes involve one or a few bases in the DNA and, apart from a small group of relatively common changes, most are vanishingly rare. This spectrum has implications for the testing strategy used for CF. Many *CFTR* mutations completely abolish function, perhaps by preventing any viable mRNA being produced, altering a part of the protein which has a critical function, or (in the case of the commonest mutation) preventing the protein product from reaching the cell surface where it is needed for ion transport. However, some mutations do not completely abolish function, and patients with one severe and one mild mutation may show a range of symptoms short of classical CF.[1]

Common Recessive Disorders

Recessive conditions might be expected to persist more in a population than dominant conditions, because most of the mutations are present in unaffected carriers, rather than those unlucky enough to have inherited two mutations. It has been shown that virtually everybody is a carrier for one or more rare recessive mutations—it only matters for those who have children with a partner who is a carrier of a mutation in the same gene, which in most cases is highly unlikely.[2] Where both parents are carriers of a mutation in the same gene, each has a 1 in 2 chance of passing that mutation on to any child. This risk is independent for both parents, so the risk of a child inheriting two mutations is 1 in 4. If it is known how common mutations are in the population, it is possible to work out the expected birth incidence of the relevant genetic disorder. If such a condition (like CF in Northern Europe) has a carrier frequency of around 1 in 25 of the population, the expected incidence of CF is calculated by multiplying the likelihood that each parent is a carrier (1/25 × 1/25) by the risk for each pregnancy that both parents pass on a mutation (1/4). For CF this gives a figure of 1 in 2,500 births. It can easily

---

[1] There seems to be a different threshold in different organ systems for adequate *CFTR* activity. For example, there are *CFTR* variants that in male compound heterozygotes (who also carry a severe mutation) can cause infertility due to absence of the vas deferens (the duct transporting sperm that is cut in a vasectomy). Other individuals with similar combinations of a severe mutation and a mild mutation/variant may have lung or pancreatic symptoms without having full CF. Yet others may have no detectable clinical symptoms at all—environmental or other genetic factors have a role. It should be stressed however that individuals with two severe mutations usually will have the full range of symptoms associated with CF. (Often in a recessive condition where there are such genotype–phenotype correlations, the extent of symptoms in an individual with a severe and a mild mutation tends to be determined by the milder of the two.).

[2] Recessive conditions may be seen more frequently in societies where relatives tend to marry, for the simple reason that if you have children with a genetic relative, it is more likely that both of you will share a rare recessive mutation, the likelihood increasing in proportion to the closeness of your relationship. There is a solid genetic grounding for the incest taboos present in virtually all human societies. The drawbacks of inbreeding might also be something worth thinking about for those who consider racial purity, whatever that might mean, to be a good thing.

be seen that for rare recessive conditions, carried by perhaps only 1 in 1000 of the population, the incidence of the disease will be very low (1 in 4 million for this example).

The population frequency of autosomal recessive conditions is not solely maintained due to the reservoir of mutations held in carriers, and therefore not directly exposed to selection. Heterozygotes for some autosomal recessive conditions gain a direct benefit from their status. This heterozygote advantage is best established for sickle-cell anaemia and for the thalassaemias, single gene conditions causing very severe blood disorders in homozygotes. However, heterozygotes for mutations in these genes have a degree of protection against malaria. These disorders are found in populations where malaria is or until recently was a problem, and collectively are the commonest severe autosomal recessive conditions worldwide. Where the advantage gained from malarial resistance in a large number of heterozygotes outweighs the clinical disadvantage incurred by a smaller number of homozygotes, the mutations are maintained in the population. It is thought that at some stage heterozygotes for *CFTR* mutations in a European population also enjoyed a heterozygote advantage, probably due to carrier status conferring some resistance to a pathogen (cholera is a likely candidate).

Where a recessive mutation arises in a small population and becomes commoner due to some selective advantage, that particular change can be maintained at a relatively high frequency in future generations derived from that original small population. This kind of founder effect is thought to be behind the fact that in Northern Europe, one particular change accounts for approximately 70% of all *CFTR* mutations (and all current carriers may be descendants of one individual in whom the mutation first arose). This mutation involves a deletion of three base pairs, and causes the removal of a single amino acid, phenylalanine, at position 508 of the CFTR protein. This mutation can be labelled in various ways but is perhaps most often referred to as delta F508, delta indicating deletion, and F being the one-letter abbreviation for phenylalanine (see Chap. 7 regarding systematic nomenclature for mutations).[3] As mentioned above, this mutation causes the failure of the protein to reach the proper place on the cell surface, and effectively completely abolishes the function of that allele. Homozygotes for delta F508 therefore have the classical symptoms of CF, but heterozygotes are unaffected, and when this

---

[3] In most cases the ethnic origin of a patient makes no difference to the choice or interpretation of a genetic test; mutations such as that causing HD are the same in all populations. There are situations, however, as in the example of delta F508, and for certain *BRCA1* and *BRCA2* mutations (see Chap. 5) where for historical reasons some mutations are common in certain populations. This might then make a difference to the most appropriate testing strategy, and on how the results are interpreted.

mutation first occurred, heterozygotes in the environment of that time had some selective advantage in comparison to non-carriers, sufficient to establish this mutation in the population (also see Box 3.1).

Box 3.1 Hereditary Haemochromatosis

Although there are *CFTR* mutations with variable effects, most of the mutations tested for, either in homozygotes or compound heterozygotes, will cause clinical symptoms. Hereditary haemochromatosis is another autosomal recessive condition, but in this case most people who inherit two mutations will not go on to develop clinical symptoms. The gene in question is the *HFE* gene on chromosome 6, and the role of the corresponding protein is in regulating absorption and storage of iron from the diet. Clinical effects stem from excess iron accumulation in a range of tissues including the liver, skin, heart, pancreas and joints. The range of systems affected is reflected in the possible clinical symptoms, among which are abdominal pain, weakness and lethargy, heart problems, cirrhosis and diabetes. Men are more likely to be affected than women are, at least in part because women are protected by blood loss (and therefore iron loss) during menstruation and pregnancy. Males tend to be affected no sooner than around 40–60 years old, and women after menopause. The environment (and almost certainly the modifying effects of other genes) also plays an important role.

The two commonest mutations are often referred to as C282Y (cysteine at amino acid position 282 in the protein replaced by tyrosine, C and Y being the one letter code for these amino acids) and H63D (histidine at position 63 replaced by aspartic acid). In both cases a single base change in the DNA is responsible. Although in Northern European populations something like 1 in 3 individuals are carriers of one of these two common mutations, only around 1 in 300 show clinical symptoms. Based purely on population frequency and an assumption of 100% penetrance, you would expect to see symptoms in more than 1 in 40 individuals. The penetrance is therefore low (less than 20% penetrance for C282Y homozygotes, less than 2% for C282Y/H63D compound heterozygotes, and negligible, if any, for H63D homozygotes).[4] *HFE* is thus an example of a gene where the line between what is a normal variant and what is a pathogenic mutation is somewhat blurred. As in the case of familial hypercholesterolaemia, introduced in Box 3.1, hereditary haemochromatosis is to some degree a disorder where clinical problems are manifest in the current conditions, but in earlier environments what are now termed mutations were likely to have been advantageous. In circumstances where life expectancy was shorter and the risk of iron deficiency due to blood loss and poor diet higher, a genotype that helps you to hang on to what iron you have would have been advantageous. With a founder effect similar to that described

---

[4] Direct-to-consumer testing for the common haemochromatosis mutations is offered by some companies. The value of this testing is debatable since the majority of those shown to have the C282Y and/or H63D mutations will never develop any clinical symptoms.

for delta F508, it is likely that the C282Y mutation only arose once, probably in a Celtic population, over 6,000 years ago. The origin of H63D is less certain, but there is evidence for a founder effect for other mutations in different populations, suggesting that the selective advantage for *HFE* gene mutations was widespread.

Testing for hereditary haemochromatosis makes use of any one of a wide variety of available methods that can be used to detect known single-base changes. The choice will depend on considerations of throughput, cost, and of how the testing fits into the local workflow. The value of testing is in confirming a diagnosis or in excluding hereditary haemochromatosis as a cause for iron overload, prompting a continued search for other causes. Where a diagnosis is confirmed a cheap and easily available treatment—phlebotomy—is available. Hereditary haemochromatosis is a disorder where applying leeches would have been a valid option (and it has been done in patients where standard methods of bloodletting were inappropriate).

Genetic Testing for CF

The testing strategy for CF is informed by the mutation spectrum: largely changes in one or a few bases, with one common mutation accounting for 70% of the total, a handful of mutations accounting for a few percentage points, and then a large number of very rare mutations.[5] Any rational testing strategy would take into account the need to capture the commonest pathogenic changes, and also be influenced by the diminishing returns due to the rarity of most of the known mutations. One possible approach would be the brute force one of sequencing the whole of the gene. This would have the advantage of capturing almost all possible changes, but is relatively expensive and slow compared to more targeted methods. Another disadvantage of full gene sequencing is that variants of uncertain significance would inevitably be detected; there is an argument that it is wise to only look for genetic changes where the significance is understood. It isn't necessarily helpful for a patient to be told that multiple variants have been detected, but the role (if any) of these variants in their clinical symptoms is uncertain.

The first line test for CF in most centres is a compromise, in that the commonest mutations in the local population are tested, which gives a reasonable sensitivity with high specificity. One of the most practical methods

---

[5] These figures are based on Northern European populations; CF is rarer in other groups, and the spectrum of mutations may differ in these groups.

involves a commercial kit which detects 50 of the commonest mutations in Northern Europe, including delta F508. Figures vary slightly but this kit detects something like 90% of mutations found in the UK. The testing of these 50 mutations uses an elegant variation on PCR, where primers are designed to be complementary to a sequence including the site of the mutation. The assay involves using two different primers for each mutation site, one primer designed so that a product is only produced if the mutation is present, and the other so that a product is only produced if the mutation is absent. By labelling the two primers with a different fluorescent tag, it is possible to distinguish homozygotes for two mutations at that site from carriers or those without the mutation (see Figs. 3.1 and 3.2). Because multiple mutations can be tested together (by designing the assay for each mutation to produce a different size of product, separable by electrophoresis), compound heterozygotes are also detected and the test is rapid and cheap. The downside is that the test will not detect the vast numbers of mutations that fall outside the regions covered. However, the trade-off seems to work well, in that at least one mutation will be detected for the majority of patients with CF. In cases where only one *CFTR* mutation is detected or where no mutations are detected and a diagnosis of CF has been established on other grounds, there is always the option of full sequencing for that sub-set of patients.

The Value of Genetic Testing in CF

A diagnosis of CF can be made on non-genetic grounds (in addition to the characteristic clinical symptoms, there are biochemical tests that can be used with a high degree of certainty for positive cases), so genetic testing for the condition needs to be justified by providing benefits other than the ability to make a diagnosis in symptomatic individuals. Genetic testing in CF is particularly useful for early diagnosis in asymptomatic children (see below) and for carrier testing, both in unaffected family members and in partners of known carriers to refine carrier risks. More recently there has arisen an increasing need to determine which mutations are present in order to establish eligibility for specific treatments, of which more below.

CF testing is now part of the neonatal screening programme in many nations. Neonatal screening involves testing small blood samples from newborn babies for a range of conditions, using biochemical and genetic tests. The rationale for this screening is to facilitate detection of conditions where early treatment or some other intervention has an important benefit for an

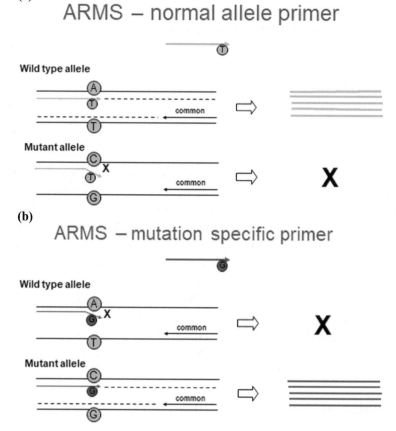

Fig. 3.1 A method commonly used to detect *CFTR* mutations. This technique is often referred to as allele-specific PCR, which sums up the purpose of the test, but is more usually referred to by the acronym ARMS (amplification refractory mutation system). Part **a** shows amplification of a normal allele, to produce amplified DNA fragments labelled with a green fluorescent dye. The primer used has been designed to be complementary to a normal allele (T forming a base pair with A in this case) but not complementary to the mutated allele (T cannot form a base pair with C and the PCR will not work). Part **b** shows the reciprocal situation with a primer specific to the mutation producing amplified fragments labelled blue. The reverse primer is common to both reactions as shown in the diagram. Diagrams courtesy of Dr Ann Curtis, Northern Genetics Service

affected individual. In the case of CF an initial biochemical test is conducted, which can indicate a raised risk of the disorder. Those with a raised risk are then tested (in the strategy used in the UK and some other healthcare

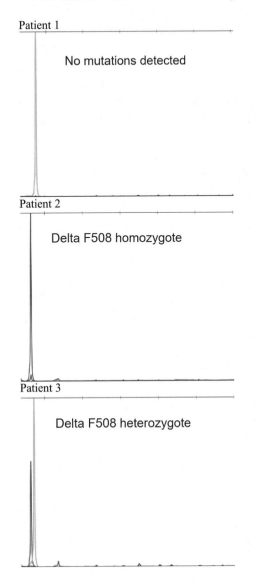

Fig. 3.2 Results of testing for 50 *CFTR* mutations in three patients. In this assay detected mutations are labelled in blue and normal alleles in green (the results for the bulk of the normal alleles detected have been omitted for clarity). In Patient 1 no blue peaks are detected, and the green peak corresponds to the normal allele at the position of the common delta F508 mutation. This patient therefore has none of the 50 mutations tested for (but note that rarer mutations are not excluded). Patient 2 has a blue peak but no green peak at the delta F508 position and is therefore homozygous for this mutation. Patient 3 (one of the parents of Patient 2) has both a green and a blue peak at this position and is therefore heterozygous for delta F508

systems) for delta F508 and three other relatively common mutations.[6] A diagnosis is confirmed in homozygotes or compound heterozygotes for any combination of these, allowing treatment to begin at a very early stage, which has been shown to improve outcomes for the affected child. In countries with such programmes the vast majority of children with CF are diagnosed in the neonatal period. CF testing for diagnostic purposes is also useful in atypical cases, where a *CFTR*-related disorder is thought possible, and the absence or presence of *CFTR* mutations can change the balance of probabilities (although because the test is not 100% sensitive uncertainties often remain, as outlined in the section on risk and uncertainty, below). Because the genotype is determined at conception, prenatal diagnosis (PND) for genetic disorders, including CF, is possible where the underlying mutations in a family are known.

Carriers for CF can only be reliably detected by genetic testing. Couples who want to know their risk of having an affected child can be offered a carrier test (although again with the caveat that not all mutations are detected by the commonly used routine tests). Where the mutations within a family are known, it is possible to definitively detect or exclude the presence of those mutations. For those with no family history, it is possible to significantly reduce the risk of their being a carrier if the common mutations in a population are excluded. Carrier testing is not generally carried out in children, on the grounds that they can give informed consent at a later stage when the information becomes relevant, for making reproductive decisions. However, this information will sometimes come to light following a diagnostic test. If PND were carried out in a pregnancy where the fetus is at a 1 in 4 risk, you would expect to find that the fetus was a carrier in half of such tests. It is not unusual for parents to request cord blood testing (a blood sample can be taken painlessly from the umbilical cord at birth) where a child is at risk of inheriting two mutations and PND is not an option they would consider. Once again, carriers will be detected 50% of the time. Ethically it is acceptable that this information can become known as an unavoidable consequence of a diagnostic or predictive test, and patients should be counselled with this in mind. Some direct-to-consumer home testing kits can also determine carrier status for common CF mutations. Confirmation of the result, familial testing and genetic counselling would then fall to the local healthcare system.

---

[6] The rationale for testing only four common mutations initially is to avoid detecting carriers for rare or mild mutations, on the basis that the vast majority of affected individuals will have at least one of these common mutations. Where only one mutation is found on the initial screen, the results for the other 46 mutations detected in the routine CF test can be examined, with the option of full sequencing where appropriate.

Prenatal Diagnosis

Genetic testing allows the possibility of prenatal diagnosis (PND), in many cases with a view to terminating an affected pregnancy. Of all of the ethical issues surrounding genetic testing, PND is perhaps the most intractable. Individuals who themselves are affected by a genetic disorder could reasonably draw the conclusion that if it is considered acceptable to terminate an affected pregnancy, society cannot be said to value people in their position, and in effect a judgement has been made that their life with a genetic condition is not worth living.[7] There are significant numbers of people for whom termination of pregnancy is unacceptable under almost all circumstances, on moral grounds (often, although not necessarily, underpinned by religious beliefs). Even among those who believe that termination of pregnancy can be justified, there is widespread uneasiness about the idea of cutting short a potential life, and about the echoes of eugenics if the assumption is made that some lives are not worth living. (Potential life is itself a loaded term which might be contested by those for whom life has already fully been established at conception.)

There isn't space in this context to do proper justice to all of the arguments on both sides, but given that genetics labs provide PND in what are regarded as appropriate circumstances (which would include full genetic counselling), it is worth outlining some of the main disputes.[8] A central issue is at what point full personhood begins. If you take the absolutist view that life begins at conception, and that the full respect owed to a human being starts at that point, then termination of pregnancy is clearly equivalent to murder. This view has the merit of clarity, but not everyone would accept that a single cell, whatever the potential, should be allowed the same rights as a fully formed person. Defining at what stage you become a fully formed person is in itself controversial. Is it pre-conception, at conception, at the time when life independent of the mother becomes possible (which in any case is continually changing with advances in care for premature babies), at birth, at a certain stage of cognitive development? Becoming a human person is a process with

---

[7] In some circumstances parents might prefer to have children with the same genetic condition that they themselves have; for example there are cases where parents with a genetic hearing loss have felt that a deaf child would fit best into their family.

[8] 'Appropriate circumstances' might include the condition that PND should only be offered for a family where there is significant risk of a severe genetic disorder being inherited. Exactly how should 'significant risk' and 'severe disorder' be defined? Neither of these judgements can be made purely objectively (although in practice risks may be quantifiable, and the option of PND usually arises for dominant or recessive single gene disorders where the risk is 1 in 2 or 1 in 4). There will be some conditions that are so severe that they clearly qualify, and some inherited characteristics that clearly do not, but also intermediate cases where different families might make a different decision.

no clear end point, and any start point which is assigned will be to some extent arbitrary.

Alongside the question of the rights of the fetus, how should these be balanced against the rights of the parents (and in this context in particular those of the mother) and of other family members? Broadly, current practice in the UK is that relevant testing laboratories should provide timely and accurate PND when the parents decide that on balance this is the best option, in the light of the family history and information provided (in a non-directive manner) by the clinical genetics team. The likelihood that parents will opt for PND may depend to a large degree on their own experience, either with their own affected child or a family member, of how much that child suffered or is suffering. It is likely that for conditions such as CF, where novel therapies will undoubtedly improve the quality of life for affected individuals (see below), the proportion of parents opting for PND will decrease.

Therapies for CF and Genetic Testing

Many advances have been made over the past few decades in proactively treating the symptoms of CF. These therapies have raised life expectancy and improved quality of life significantly for patients, but until recently a specific treatment of the underlying genetic cause has been lacking. Much of the early work on gene therapy centred on CF, as a relatively common (in some populations), severe and well-understood condition. For many years it has been hoped that gene therapy, via a targeted replacement of mutated sequences or by delivering DNA or RNA to produce an absent product, could cure genetic disorders. It is fair to say that in most cases these hopes have not been realised (although there are exceptions, and some of the techniques involved have also been put to stunningly successful use in the novel COVID-19 vaccines). However, in the last few years a range of drugs has become available, designed specifically to mitigate the damage done to CFTR function by mutations. These CFTR modulators can improve function in a range of ways, including helping to stabilise the protein, correcting the protein shape, ensuring that the protein reaches the cell membrane, and helping to increase levels of functional protein. Different combinations of drugs can be appropriate depending on the underlying mutations present in that patient—truly personalised medicine. Correct and timely determination of individual genotypes can therefore be essential to ensure that eligible patients are treated. Although at present this isn't suitable for all patients, the availability of these drugs used in combination means that potentially most CF patients can be treated in a far more effective way than just by managing their symptoms.

Calculating Risks

Unlike HD, where the diagnosis can almost always be made or excluded with certainty by a genetic test, the spectrum of mutations seen in CF leads to scenarios where no absolute answer is possible for the patient, but rather a change in probabilities. In the case of single gene disorders, it is often possible to know the initial probability that an individual is (for example) a carrier of a mutation based on their family history. The results of testing can be used to give a more accurate risk, even if not a certainty.

Scenario 1

Suppose you have one parent who is a carrier for an autosomal recessive disorder like CF; your risk of inheriting that mutation is 1 in 2. Ideally it would be possible to detect the mutation in the parent, and then perform a definitive test to show whether or not you have inherited that change. This isn't always achievable: your parent may be an obligate carrier based on family history but not willing to give a sample to be tested, or may have died before giving a sample, or may be uncontactable for whatever reason. Your risk can still be considerably reduced if you are tested for the common CF mutations and none is present (this still leaves a risk that your parent was a carrier of a rare mutation that they have passed on undetected).

Scenario 2

Commonly where one of a couple has been shown to be a carrier of CF (usually on the basis of a family history where the affected individual has been tested and has known mutations) the couple will want to know the risk of having an affected child, which will depend upon whether or not the other partner is a carrier. If the partner is assumed to be at population carrier risk (having no family history) this risk would vary depending on the figures for the population in question (1 in 25 in Northern Europe) but again this risk could be reduced by excluding the presence of the most common CF mutations.

Intuitively, and in the absence of any family history, it seems to make perfect sense that if you don't carry any common mutations, you are less likely to be a carrier of CF. It is possible to quantify this reduction in risk using calculations based on Bayes' theorem.[9] Bayesian statistics are used in a whole range of fields, but in genetics this approach is used to modify the initial likelihood of a state of affairs such as being a carrier of CF (known as the prior probability) in the light of additional information such as the results of genetic testing. Multiple pieces of information can be used as long as they are independent of each other. The resulting final probability will be approximate (involving assumptions about population carrier frequency and/or the proportion of mutations detected in that population for the scenarios above) but will give patients and clinicians a more accurate picture.[10] For the mathematically minded there are numerous full explanations available online, but for the examples above in scenario 1 (prior probability of being a carrier 1 in 2, familial mutations unknown, and assuming 90% of mutations excluded) the final carrier risk would be approximately 1 in 11. For scenario 2 (population risk 1 in 25) the final carrier risk would be approximately 1 in 240. These changes would make a significant difference to the overall likelihood of having an affected child.

Giving results in terms of probabilities is one of the great challenges for genetic counselling. It is well established that the perception of risk varies widely from individual to individual, and an optimist might react quite differently from a pessimist on being given the same risk. The way that risks are framed also affects our response; tests have been rated quite differently in terms of approval when stated to be 90% correct than when stated to make errors 10% of the time, even though the figures say the same thing.

---

[9] Bayes' theorem is named for its discoverer, the Reverend Thomas Bayes, an eighteenth-century clergyman and mathematician who it is safe to say couldn't possibly have envisaged this use of his method.

[10] It is essential that the assumptions made are clearly stated. It is unhelpful (and dishonest) to present results as more certain than they are or to give the impression of a greater degree of accuracy that is warranted by the nature of the test.

4

X-linked Inheritance: A Question of Gender

Summary The implications of having a different chromosome complement in males and females are explored in this chapter, with particular reference to X chromosome genes involved in genetic disorders. The epigenetic mechanism for inactivation of an X chromosome in female cells is discussed. Three examples are given of genes on the X chromosome that illustrate some of the consequences of the XX/XY basis of sex determination. All three of these genes (the *DMD*, *FMR1* and *AR* genes) also show how different mutations in the same gene can give rise to different phenotypes. The utility of testing for these different conditions and some of the genetic counselling aspects are covered briefly. The technical basics of the methods used to detect these different mutations (and of a method for determining patterns of X-inactivation) are introduced. The final part of the chapter discusses to what extent a genetic viewpoint is useful in clarifying wider questions of gender.

A Fundamental Imbalance

Most genes are present in equal numbers in all humans. The obvious exceptions are those on the X and Y chromosomes; those on the X are present in a double dose in XX females and those on the Y are present only in XY males. How does this imbalance work, given that in many instances loss of a functioning copy of a gene can have significant effects? In the case of the Y chromosome, the number of genes is low—perhaps around 100 functioning genes of the total of some 20,000 (depending on the exact definition of gene used). Over the course of evolution the Y chromosome has lost many genes, and clearly it is possible to get along perfectly well without those genes only found on the Y chromosome.[1] The situation regarding the X-chromosome is more complex. There are of the order of 1,000 genes on the X, and it is difficult to see how this level of imbalance in the copy number of genes could be accommodated. For most chromosomes, lacking a copy or having an extra copy is incompatible with life.[2] It has to be the case that (just as it is possible to get along without any Y chromosome) it is possible to manage with only one X. That being the situation, how is the double dose of X chromosomes in each female cell sustained? The explanation lies in X-inactivation, a process which ensures that only one X is actually active in every cell containing more than one X chromosome.

Switching Off Genes on the Inactive X Chromosome

One of the (many) control mechanisms for gene expression is methylation of DNA. A methyl group (chemically CH_3, a carbon atom with three hydrogen atoms) is attached to the DNA base cytosine to form 5-methycytosine (the 5 refers to the position of attachment of the CH_3 molecule). In humans, this occurs in regions rich in the bases C and G, which tend to occur in the control regions of genes just upstream of the start of the coding sequence (referred to as the promoter region of the gene). Where the promoter region is methylated genes are effectively switched off, as they become inaccessible to the signals that would otherwise cause gene expression to be activated; methylation is associated with chromosomes becoming more condensed, more tightly packed. Methylation is therefore an example of an epigenetic

---

[1] The remaining Y-specific genes include the *SRY* gene (sex-determining region of the Y chromosome), which is responsible for the cascade of events leading to development as a male, along with other genes related to male development and fertility. The situation as described above ignores the fact that there is a small region of the X and Y chromosomes that is shared, containing some genes present in two copies in both females and males (the pseudoautosomal region). However, the vast majority of X-specific genes are present in two copies in females and one in males, and most genes on the Y chromosome are present in males only.

[2] Aneuploidies, instances of an abnormal number of chromosomes in a cell, involve such huge imbalances in gene activity that those affected with the loss or gain of a whole chromosome don't usually make it to birth. The main exception, (apart from the special case of the X and Y chromosomes) is trisomy 21 (three copies of chromosome 21), more usually known as Down syndrome. Chromosome 21 is the smallest autosome with the lowest number of genes, but the extra copy still has profound effects. The normal chromosome count (44 autosomes in 22 pairs plus XX or XY) is described as diploid. Cells with only one copy of each autosome plus X or Y (such as sperm or egg cells pre-fertilisation) are described as haploid. Some cancer cells have grossly abnormal chromosome complements, with three or more copies of all chromosomes. Such cells are described as polyploid.

process, one in which gene expression is controlled by factors beyond the DNA sequence (also see Chap. 7). Once the techniques of cell preparation and microscopy made it possible to distinguish human chromosomes, it was observed that in female cells with two X chromosomes one of the X chromosomes was always much more densely packed than the other. This is the inactive X. (In females with three X chromosomes, which usually causes few or no apparent symptoms, there are two inactive X chromosomes in each cell.) Subsequently, it was shown that this inactivation is secondary to methylation of (almost) the whole length of the X chromosome. The problem of imbalance is solved by using methylation to switch off genes on the inactive X, leaving genes to be expressed only from one X per cell, as is the case in males.[3]

The 'choice' of which X chromosome is shut off appears to be random (except where there is a selective reason to switch off one X rather than the other, as is the case where one X is damaged in some way). However, once that choice has been made, all daughter cells will have the same X inactivated. X-inactivation occurs during embryonic development, and the random nature of the switch means that different cells will make a different decision. There will therefore be different cell lineages in every female in regard to which X is active; all women are mosaics in this respect (mosaicism being the general term for the co-existence of two or more genotypes within a single individual). The stochastic nature of the process, and the fact that choices are made at a time in development when there are relatively few cells, together imply that by chance one of the two X chromosomes may be predominantly active rather than it being 50:50 across all tissues. This possibility of skewing of X-inactivation, either by chance or because of damage to one X leading to preferential inactivation of the damaged X as a protective response, has implications for X-linked inheritance in females. This will be explored further in the context of specific disorders.

Inheritance of X-linked Genetic Disorders

The chromosomal basis of sex determination in humans (XX females, XY males) is not universal among complex life: in birds males have the same sex chromosomes (ZZ in this case) and females two different chromosomes (ZW). There is no reason why sex determination can't just be based on the

---

[3] This is something of an oversimplification, as among other things some genes on the inactivated X do escape inactivation, but it is essentially the case that any imbalances are avoided by switching off genes where a double dose would be harmful.

ratio of X chromosome to autosomal genes and this is the case for many animals; in insects for example, males are X0 (meaning just having an X chromosome) and females XX. Given that we do have the XX/XY system, certain consequences follow.[4] The most obvious consequence is that X-linked genetic conditions usually only affect males or affect males more severely. For X-linked recessive conditions, female carriers may be unaffected (but see below for exceptions) while for X-linked dominant conditions females may be more mildly affected than males. Perhaps surprisingly, for some severe X-linked conditions only affected females are seen. This apparently counterintuitive finding is explained by the fact that these conditions are so severe in males that they are lethal in utero, so affected males don't make it as far as birth. The following sections give some specific examples of how X-linked inheritance can affect patients in practice.

Three X-linked Genes Associated with Genetic Disorders

The following part of the chapter will briefly introduce three very different X-linked genes and the associated disorders. These examples demonstrate the effects of a variety of types of mutation that cause the genetic conditions, as well as some of the consequences of an X-linked inheritance pattern. The three genes share the feature that, in each case, alternative mutations within the same gene can give very different clinical effects—examples of genotype–phenotype correlations.

---

[4] Women in general have longer life spans and better overall health than men. There will be many factors contributing to this, among which is the possession of two X chromosomes, allowing for the possibility that any damaging mutations in a gene on one X chromosome can be compensated for by a functional copy on the other X. This is an option not available to a male, where any sub-optimal X chromosome genes will be exposed. Although in each individual female cell one X chromosome is inactivated, in the absence of significant skewing, both X chromosomes will be active in different tissues, giving the potential for extra resilience. It may also be the case that for any phenotype determined in part by an X-linked gene or genes, males are more likely to be at extreme ends of the spectrum for that particular feature; given that there will be no mitigation from a second copy of those genes.

X-linked Example 1: The *DMD* Gene and Duchenne Muscular Dystrophy/Becker Muscular Dystrophy

The product of the *DMD* gene (dystrophin) is a protein that has a critical role in muscle structure.[5] Complete or partial loss of this protein leads to severe muscle wasting. Boys with Duchenne muscular dystrophy (DMD) are usually wheelchair bound by their teens, and life expectancy is severely reduced, due in most cases to respiratory problems but also to cardiac dysfunction. Mild cognitive impairment may also be a feature. Boys with Becker muscular dystrophy (BMD) are less severely affected and in most cases female carriers of *DMD* mutations do not show clinical symptoms. The functioning copy on the second X-chromosome is usually sufficient to maintain normal function (but see below for some exceptions). Collectively, dystrophinopathies (any clinical phenotype caused by a *DMD* gene mutation) are among the commonest severe genetic conditions, with a birth incidence in males of around 1 in 3000 to 1 in 5000. The relatively high incidence is in part explained by the presence of a pool of mutations in unaffected female carriers, and in part by the frequent occurrence of new mutations.

The Spectrum and Significance of Mutations in the *DMD* Gene

The *DMD* gene is one of the largest human genes known (spanning more than 2 million bases from the promoter regions to the end of the gene).[6] The gene has 79 exons, and because this region of the X chromosome is particularly susceptible to loss or gain of DNA (deletions or duplications) most of the mutations (around 80%) involve a deletion or a duplication of one or more of these exons. It is important to know exactly which exons are duplicated or deleted as the nature of the mutation can determine whether a patient has DMD or the milder BMD. The concept of frameshift mutations was introduced in Chap. 1, in the context of loss or gain of a small number of bases (a frameshift occurring where the change doesn't involve a multiple of three bases which would maintain the reading frame). The same principle

---

[5] There is potential for confusion between the dystrophin gene name (*DMD*, always italicised) and the abbreviation for Duchenne muscular dystrophy (DMD, not italicised).

[6] There are several alternative promoters for the *DMD* gene, producing different versions of the dystrophin protein in different tissues. This is a common pattern for human genes, allowing far more proteins to be produced than might be assumed from the total number of genes.

applies on a larger scale for losing or gaining one or more exons—this can be in-frame or out of frame (other than at the start and end of the coding sequence, an exon can start and finish at any of the three positions of a codon; see Fig. 4.1). In order for the coding sequence to be in-frame, the final base of an exon and the first base of the next exon must occupy successive positions in a codon. There are only three possibilities: if the exon ends in the first base of a codon, the next exon must start with the second base, exons ending in the second base of a codon are followed by exons starting with the third base, and if an exon ends with the final base of a codon the next exon starts with the first base of the next codon.

If a deletion or duplication is out of frame, and therefore no functional protein is produced, this is (usually, there are exceptions) associated with the more severe DMD. In-frame deletions or duplications are usually associated with the milder BMD, reflecting the fact that some protein function can be maintained. The severity of the phenotype often correlates better with the frame of the deletion or duplication than with the size; some multi-exon in-frame deletions have a milder phenotype than single exon out of frame deletions (although this does also depend on exactly which exons are involved). This distinction also applies to *DMD* gene mutations involving one or a few bases: a nonsense mutation would usually be associated with

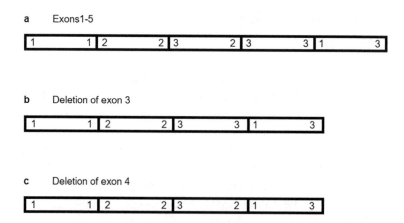

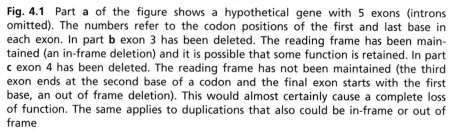

Fig. 4.1 Part **a** of the figure shows a hypothetical gene with 5 exons (introns omitted). The numbers refer to the codon positions of the first and last base in each exon. In part **b** exon 3 has been deleted. The reading frame has been maintained (an in-frame deletion) and it is possible that some function is retained. In part **c** exon 4 has been deleted. The reading frame has not been maintained (the third exon ends at the second base of a codon and the final exon starts with the first base, an out of frame deletion). This would almost certainly cause a complete loss of function. The same applies to duplications that also could be in-frame or out of frame

DMD, and a missense mutation that allows some retention of function with BMD.

The instability of the genomic region containing the *DMD* gene leads to the frequent occurrence of deletions and duplications as de novo (new) mutations. In about one-third of cases, the mother of a boy with DMD will not herself have inherited that mutation (the mutation is not detectable in a blood sample and is therefore not present in all of her somatic cells, and so cannot have been passed on from a parent).[7] Where a mutation is inherited by a woman this is usually maternal in origin, but if this individual's father had a *DMD* gene mutation, and if he was sufficiently mildly affected so that he was able to have children, a carrier could have inherited a mutation paternally. Males with an X-linked mutation will pass this on to all of their daughters (who will therefore be obligate carriers) and to none of their sons.

This section so far has only considered the effect on female carriers in terms of risk to children. However, women who carry a mutation are also themselves at some risk of clinical symptoms. Some carriers develop age-dependent cardiomyopathy, and muscle weakness may also occur. The degree to which women are manifesting carriers may depend on patterns of X-inactivation. As mentioned, it is possible that just by chance, the same X-chromosome might be inactivated in most tissues (and skewing tends to get more pronounced with age). If the inactive X happens to be the one with the normal *DMD* gene copy, the active X in most tissues would not have a functional *DMD* gene, and clinical symptoms would be more likely or more severe. This wouldn't necessarily have to be a global effect: depending on which X-linked gene was involved, skewing of X-inactivation in one critical tissue might be enough for clinical problems to develop. There are also circumstances where skewing is favoured for other reasons than chance, such as scenarios where a chromosomal rearrangement has occurred such that cells which inactivate a specific X are not viable, so selection will ensure that the same X is active in all tissues.[8] Once again, this could lead to expression of an X-linked phenotype. Another way in which females can manifest an X-linked recessive condition

---

[7] The de novo change will have occurred either in the individual egg which gave rise to her affected son, or at some stage in the development of her germline cells. This is a crucial distinction as in the first instance there would be a negligible risk of having another affected child, whereas if her egg-producing cells contain the mutation on one X chromosome it will be present in half of her eggs, and the chance of a male child being affected is 1 in 2 (and any daughters have a 50% chance of being carriers).

[8] Prior to the availability of genome-scale sequencing, finding the position of genes including the *DMD* gene often depended on very rare events such as chromosomal rearrangements which disrupted genes, causing the clinical phenotype, and the location of the rearrangement by chromosome analysis then helped to localise the gene.

is in Turner syndrome, where one X chromosome is absent.[9] As only one X chromosome is present, this will be active in all cells. It follows that if an individual has Turner syndrome and an X-linked mutation on their only X chromosome they are likely to be affected.

Genetic counselling for affected families is of great importance to explain the various complexities outlined above. Counsellors may have to deal with particularly difficult situations where (for a variety of reasons) individuals are unhappy to share information that could be critical to allow relatives to make decisions (for example on carrier testing and prenatal diagnosis). Once again, the question arises as to whether genetic information should be solely held by the individual tested, or whether this should be considered joint information for the family.

Testing for *DMD* Gene Mutations

Finding a *DMD* mutation is of great importance in confirming a diagnosis of DMD/BMD or in establishing carrier status. Because the great majority of mutations are deletions or duplications, most methods have been aimed at this type of change. Detecting deletions in affected males is the most technically straightforward, and was the first widespread direct mutation test for DMD and BMD. The approach used was multiplex PCR (amplifying multiple DNA fragments in the same reaction tube by using multiple sets of primer pairs for different exons of the *DMD* gene). If a male patient has a deletion of an exon or exons in the *DMD* gene, in the absence of a second X there is no target for the PCR primers, and therefore no amplification. The inclusion of primer pairs for multiple exons ensures that some will be amplified, and so confirm that the reaction has worked properly and that amplifiable patient DNA was present. Each different PCR primer pair is designed to give a different size amplified DNA fragment, separable by electrophoresis. If a fragment or fragments are absent, this shows that the corresponding exons are deleted in the patient (see Fig. 4.2).

This method is cheap and straightforward, but limited in scope. If a deletion is found in an affected male, carrier testing in female relatives would be the obvious next step in what is usually referred to as cascade

---

[9] Other than the disposability of the Y chromosome, this is the only example where lack of an entire chromosome is compatible with life, reflecting the unique situation with respect to the X chromosome. Turner syndrome causes problems with fertility and some other health issues, but in many cases doesn't prevent individuals having a relatively normal and healthy life.

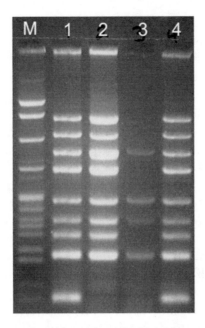

Fig. 4.2 Multiplex PCR-amplified fragments from the *DMD* gene in male patients. The fragments have been separated by electrophoresis, in this case using a gel made from agarose, incorporating a dye that binds to the DNA and makes the fragments detectable by photography over UV light. Smaller fragments migrate further (towards the bottom of the figure). The gel lane labelled M has been loaded with control markers of known size. Lanes 1 and 4 contain samples from patients without deletions. Lanes 2 and 3 contain samples from patients with different deletions, encompassing different exons. In Lane 2 only the smallest amplified fragment is absent (which would require independent confirmation to exclude a technical artefact). In Lane 3 five fragments are absent, corresponding to a deletion of five exons. However, the quality of the results for lane 3 are poor (the amplified fragments are faint) and the test might need to be repeated for this patient for confirmation. Photograph courtesy of Dr Ann Curtis, Northern Genetics Service

testing.[10] However, detection of a heterozygous *DMD* gene deletion is not usually possible using standard PCR because the method is not quantitative. The final amount of product for a PCR amplification (which can easily be measured using fluorescent PCR and capillary electrophoresis) does not necessarily reflect the amount of starting material. (In Fig. 4.2 differences in the amount of product for the different amplified fragments can be seen within as well as between patient samples.) A female carrier of a deletion has a normal undeleted X chromosome, so the PCR primers for the deleted exon or

---

[10] Cascade testing is a general term referring in genetics to the fact that a diagnosis often has immediate implications for other family members, and diagnostic or carrier testing is then cascaded through the family.

exons will still find a target on the normal X, and the final amount of product does not distinguish between having one or two copies of that target per cell. The same applies for males with a duplication or female carriers of a duplication; a PCR will not distinguish between the normal one copy in a male or two copies in the case of a duplication (female duplication carriers will have a total of three copies). It is possible to get round this limitation using some tweaks in the way that PCR is set up, but a more elegant solution came with the invention of multiplex ligation-dependent probe amplification (mercifully shortened to MLPA). The technical basis of this can be found (among many other resources) on the website of the company that developed and marketed the technique (MRC Holland, see www.mrcholland.com). Essentially, what the technique allows is to determine the number of copies per cell of different genetic regions. This is done by measuring the ratios between amplified DNA fragments corresponding to the gene of interest, and amplified control regions. If control fragments of known copy number are tested at the same time as the gene of interest, for any exon of that gene it is possible to say whether that stretch of DNA is present at 0, 1, 2, 3 (or more) copies per cell, and therefore to detect deletions and duplications. A refinement of the MLPA technique enables the simultaneous testing of multiple gene and control fragments (although because the *DMD* gene has 79 exons two different reactions have to be used to test all of them). MLPA therefore allows the rapid and cheap detection of a deletion or duplication of any combination of these 79 exons. Determining precisely which exons are deleted or duplicated, and therefore whether these mutations are in-frame or out of frame, is not only important in indicating a diagnosis of DMD or BMD. This also allows access to new targeted therapies that are specific for certain mutations, analogous to the situation described for CF.

Although deletions and duplications only account for approximately 80% of *DMD* gene mutations, so a diagnosis cannot always be made or excluded using this technique alone, MLPA is an excellent first line test that in most cases only detects changes of known significance. Where a diagnosis is still suspected and no deletion or duplication has been found, full gene sequencing will detect most of the remaining mutations, so diagnostic testing, carrier testing and (if requested) prenatal diagnosis are available for almost all families.

X-linked Example 2: The *FMR1* Gene, Fragile X Syndrome and Other Phenotypes

The *FMR1* gene is similar to the *HTT* gene in that a trinucleotide repeat region is prone to expansion. The similarities, however, end there. In the case of the *FMR1* gene, part of the CG-rich promoter region contains a repeat array of the sequence CGG. This is not a coding part of the gene, so does not produce a string of identical amino acids (unlike the repeat in the *HTT* gene), but is part of the upstream control region. The normal range of repeats goes up to 58 (some sources would quote a slightly different range, and there are the usual grey areas). Repeats between 59 and 200 are classed as premutations.[11] Expansions of more than 200 repeats are classed as full mutations, and these have a completely different effect from the gain of function changes seen in HD. Once a repeat gets to around the 200 level (clearly this is a suspiciously round number—the exact threshold is uncertain, but is somewhere around this point) the C bases in the CG rich promoter region gain methyl groups and the promoter region undergoes hypermethylation which has the effect of switching off the gene (another example of an epigenetic effect). An *FMR1* allele with a full mutation is effectively hypermethylated and switched off in the same way as an *FMR1* allele on an inactive X chromosome. Large repeats in this gene therefore cause a loss of function mutation, in contrast with the gain of function seen in HD. The FMR1 protein produced by this gene is needed for typical brain development. The loss of this protein causes fragile X syndrome, the most important clinical features of which are cognitive and behavioural problems.[12] This condition represents the commonest cause of X-linked cognitive impairment with an incidence of around 1 in 4,000 males (fewer females are affected, in line with the mode of inheritance). There are other very rare mutations that cause fragile X syndrome (as might be expected, given that this is a loss of function mutation). However, the vast majority of cases are due to the CGG expansion. The name of the condition is derived from the observation that in chromosome preparations from affected patients an apparent break on the X chromosome (a fragile site) can sometimes occur due to the presence of a large expansion.

---

[11] Premutations are prone to expansion on transmission, and the likelihood of expansion increases with the size of the repeat, and with the purity of the repeat; typically the CGG repeats are periodically interrupted with the sequence AGG in normal alleles, which has a stabilising effect.

[12] A syndrome in the context of genetic disorders meaning a collection of clinical effects that (generally) occur together and are ultimately caused by a mutation in a specific gene or genes.

Multiple Conditions Are Associated with *FMR1* Gene Mutations

In the case of BMD and DMD the difference in phenotypes caused by different mutations is significant, but is one of degree rather than of kind. In the case of fragile X, entirely different conditions are caused by different genotypes. Most importantly there is full fragile X syndrome, caused by hypermethylation of the promoter region secondary to a CGG repeat expanding to more than 200 repeats. A male with this genotype will usually have the classical features of the syndrome. Females can be affected to varying degrees and may be unaffected.[13] This is likely to be in large part due to the chance effects of X-inactivation. It may be that in a crucial tissue the damaged *FMR1* gene, having been switched off by hypermethylation due to an expansion, is predominantly or wholly on the active X.[14] The counterpart allele is switched off because it is on the inactive X. The result would be a tissue with no expression of the gene, effectively the same situation as in a male with the same expansion. More severe clinical effects would therefore be expected than in a female with the damaged *FMR1* gene predominantly on the inactive X, leaving the functional copy to be expressed. Predicting how a female child might be affected (when tested by prenatal diagnosis or through early testing due to a family history) is therefore very difficult.

The other clinical problems associated with the *FMR1* gene are due to premutations rather than the full mutation, and therefore have a completely different causal mechanism, still not well understood (but probably involving a toxic gain of function of an abnormal mRNA). Female premutation carriers may undergo early menopause (fragile X-associated premature ovarian insufficiency, FXPOI) and both females and (more frequently) males with a premutation may develop a neurodegenerative condition in later life (fragile X associated tremor/ataxia syndrome, FXTAS).

---

[13] In practice it might be difficult to say in any individual case whether (for example) IQ was reduced by a small percentage compared to what might be expected for that particular individual if they were not heterozygous for an *FMR1* full mutation.

[14] There are laboratory tests for measuring skewing of X-inactivation (see below in the section on the androgen receptor gene). However, these only reflect the pattern of X-inactivation in the tissue from which the DNA is derived, and are usually carried out on DNA prepared from blood cells. This accessible tissue may not be the best one to test for a condition where gene expression in the brain is of central importance.

Transmission of Fragile X Syndrome

The relatively common nature of fragile X syndrome can be explained by the tendency of the CGG region to expand, and the presence of a pool of unstable premutations in the population. Remarkably, it is the case that large expansions are seen exclusively on female transmission. Males with a premutation will pass on only that premutation to their daughters, who will all be obligate carriers. The daughters will then be at risk of having children with a full mutation if the premutation expands on transmission to the next generation. This leads to complex patterns of inheritance in fragile X families, with the disorder apparently skipping a generation when passed on via a male with a premutation.

Testing for *FMR1* Gene Mutations

Testing is carried out using a combination of methods previously described (fluorescent PCR and capillary electrophoresis) although the PCR-based methods used are somewhat more sophisticated than those needed for HD testing. This is because *FMR1* expansions are much larger, and the region of interest is very GC rich. PCR works less well for large fragments and regions that have more C-G base pairs than A-T base pairs. In order to capture all expansions, the most widely used test combines PCR conditions optimised to produce large fragments, (as for *HTT* expansions, the PCR primers flank the repeat region) with a tweak on basic PCR using additional primers that effectively produce a ladder of fragments representing every possible repeat length up to the largest. (A similar pattern can be seen in Fig. 2.2 for the patients with an *HTT* gene expansion, although not generated in exactly the same way.) Essentially the same method can be used whether the referral is to confirm or exclude a diagnosis of fragile X syndrome (which can be done to a high degree of certainty), a carrier test, or a test for FXPOI/FXTAS. Additional testing can be done to detect hypermethylation of the *FMR1* gene promoter region where relevant.

X-linked Example 3: The Androgen Receptor (*AR*) Gene, Spinal and Bulbar Muscular Atrophy and Androgen Insensitivity Syndrome

The *AR* gene codes for a protein that interacts with testosterone and other androgenic hormones (androgenic hormones are responsible for the development of male characteristics). After binding with these hormones the androgen receptor switches on genes that are part of the cascade of changes needed for standard male development. Mutations in the *AR* gene are involved with two very different phenotypes, spinal and bulbar muscular atrophy (SBMA, sometimes referred to as Kennedy disease) and androgen insensitivity syndrome (AIS). These conditions are much more rare than the conditions described previously; SBMA is present in fewer than 1 in 100,000 males and population studies suggest that AIS in its various forms, although more common than SBMA, is still extremely rare.

SBMA: A Trinucleotide Expansion Disorder

The neuromuscular disorder SBMA is caused solely by expansion of a CAG repeat region in the coding part of the *AR* gene—a gain of function mutation as seen in HD. The threshold for clinical symptoms (which include adult onset muscle weakness and atrophy) is also very similar to that seen for HD. An individual with up to 34 CAG repeats will show no clinical symptoms whereas individuals with over 38 repeats will be affected. Larger repeats are correlated with earlier onset and alleles of 36–37 repeats may show reduced penetrance. (The comparative rarity of SBMA has meant that less is known about the effect of alleles at the top end of the normal range and the lower end of the affected range; the status of alleles of 35 repeats is uncertain.)

However, SBMA differs crucially from HD in that males and females are not equally affected. Female carriers are usually unaffected or at worst mildly affected. At first sight this seems counterintuitive for a gain of function mutation. The explanation is that female carriers are thought to be protected by having low levels of circulating testosterone and other androgens that stimulate the androgen receptor.[15] The dynamics of the CAG repeat in the *AR* gene also differs from that in the *HTT* gene in that increases in expansion size are only rarely seen; the disorder is maintained in the population (at a

---

[15] Genes that only show an effect in one sex due to their role are termed sex-limited. Examples would include genes involved in ovarian development in women, or genes controlling prostate development in men.

much lower level than HD) because of a reservoir of expansions in female carriers rather than through de novo expansions. Any daughter of an affected male will be an obligate carrier. Testing for SBMA is essentially identical to HD testing, using fluorescent PCR with primers flanking the repeat region. Confirmation or exclusion of a diagnosis or of carrier status can be made with close to 100% certainty.

The assay used for testing the CAG repeat to investigate SBMA, with certain modifications, can also be used to detect skewing of X-inactivation in patients referred for other reasons. An enzyme which distinguishes between methylated (inactive X) and unmethylated (active X) DNA is used to treat the patient sample in a way that prevents PCR amplification of the allele on the active X. In a female heterozygous for different CAG repeat lengths the patterns of amplification can be compared from untreated and treated DNA. If the patterns differ sufficiently this indicates a deviation from random X-inactivation (see Fig. 4.3).

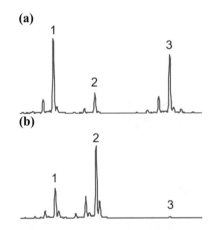

Fig. 4.3 Part **a** shows PCR amplification of the two *AR* gene CAG repeats from a female patient (untreated DNA, peaks labelled 1 and 2). The height difference between the two peaks in this case is a typical feature of PCR, in that smaller fragments (seen towards the left of the figure) tend to amplify more efficiently. Part **b** shows PCR amplification from the same patient after treating the DNA with an enzyme that prevents amplification from the active X. The peak labelled 3 in both parts derives from a male control DNA sample, used to show that the enzyme treatment has worked, since in a male the only X chromosome is always active, and therefore no (or very little) amplification should be seen after treatment. If X-inactivation were random in this tissue, parts **a** and **b** would show similar results for peaks 1 and 2. In this case the pattern differs, and one peak (peak 2 which was relatively low in the untreated amplified DNA) predominates in the treated DNA, indicating that the X chromosome from which this allele is derived is predominantly inactivated

AIS: Loss of Function Mutations in the *AR* Gene

AIS is suspected in individuals who have an X and a Y chromosome, but have some combination of atypical development of the external genitalia (which may be seen at birth), abnormal development in puberty, and infertility. The condition covers an entire spectrum from complete AIS where the external genitalia are apparently typically female, though partial AIS with ambiguous genitalia, to mild AIS with typically male external genitalia. The extent of the effect is at least in some families genetically determined, particularly at the complete end of the spectrum. Some mutations lead to a complete loss of function, while with others the function is impaired to varying degrees, which in turn determines the extent to which typical male development is affected: loss of the androgen receptor protein means loss of response to the androgenic hormones such as testosterone. Female (XX) carriers usually do not show any symptoms. This is one of many disorders that can lead to an individual having intermediate sexual characteristics. The *AR* gene is thus another example of a gene where different types of mutations (in this case gain of function and loss of function) can give rise to entirely different effects (although there is some overlap in that male patients with SBMA do show signs of androgen insensitivity, including reduced fertility, testicular atrophy and swelling of breast tissue). More than 500 different presumed loss of function mutations have been identified in the *AR* gene, and sequencing of the entire gene by Sanger or next-generation sequencing is the usual diagnostic approach.

Genetics and Gender

Biological sex determination and social determination of gender are increasingly fraught topics. The historical tendency has been to take the binary nature of XX/XY and impose this dichotomy on all of humanity. This lack of nuance, while having the benefit of clarity, does not reflect a much more complex real-world situation. Biological characteristics are usually on a continuous spectrum. This is easily understood for features such as height; in practice it will always be arbitrary to draw an exact line where someone becomes defined as tall (or short).[16] For most complex human behavioural characteristics, it would seem overly constrictive to divide everyone into one

---

[16] Height is to a significant extent genetically determined: given adequate nutrition, height is highly heritable. Heritability in this sense has a specific technical meaning, roughly speaking the degree to which genes determine the variation within a population for a given environment.

of two stable and permanent groups, rather than placing individuals at some point on a spectrum that might change with circumstances. However, the default position for gender is to divide all humanity rigidly into two groups.

On the face of it, there is an apparent genetic justification for this approach in that biological sex determination does start with a binary presence/absence of the *SRY* gene, the Y-chromosome gene responsible for starting the complex cascade of events leading to a male phenotype. In the absence of a Y chromosome, and therefore the absence of the *SRY* gene, development will be as a female.[17] However, the chain of events after this point becomes increasingly complex, with the effects of multiple genes and environmental factors interacting. Problems anywhere in this cascade of events (as seen above for AIS) can lead to what are termed collectively disorders of sex development, and once again a whole spectrum of characteristics can be seen. Even looking narrowly at purely biological factors there is a blurring of the distinction between female and male, although most individuals will be recognisably close to one end or other of the spectrum. Once complex social and psychological factors are factored in, an unambiguous binary picture seems even less plausible. It is not unusual to hear of a particular sexual orientation being condemned as 'unnatural'. This description fails on a literal reading as pretty much any combination can be seen in nature. 'Unnatural' could be more plausibly defined on evolutionary grounds as a genetically determined behaviour that decreases the likelihood of reproduction.[18] However, this takes too narrow a view on what allows a particular version of a gene to persist and thrive. Alleles increase in frequency where within a population, on average, an individual is more likely to pass on that allele than a rival allele. It is the performance of that allele in all sorts of combinations that determines its success or otherwise, so any simplistic suggestion that an allele that predisposes (for example) to homosexuality would be selected against, ignores the fact that such a hypothetical allele would have different effects on a different genetic background and in a different environment. (In any event, it is highly unlikely that any single variant in itself would have more than a marginal effect on complex aspects of behaviour.) Although the genome clearly has a central role in biological development, behaviours are formed by a complex interaction between genes and environment. Even to the

---

[17] As ever there are exceptions: some individuals with male characteristics have two X chromosomes. This may be due to a rearrangement of the genetic material in these individuals. In these rare cases the *SRY* gene has been moved to another chromosome, usually the X chromosome, and is therefore present in the absence of a full Y chromosome.

[18] On such grounds the social practices of contraception or voluntary celibacy could be judged as unnatural as it is possible to be, although both are widely considered as desirable in the right context.

extent that the genome in isolation determines sexual development, the end result doesn't necessarily fit into simple categories, and which behaviours are deemed normal is dependent to a large degree on transient social assumptions and biases.

5

Genetic Testing in Cancer

Summary This chapter introduces the concept of cancer as a genetic disease, and the consequent need for genetic testing to help guide diagnosis and treatment. A distinction is made in this chapter between acquired cancer and cancer due to an inherited predisposition, but both are equally genetic in origin (and in practice there are, as in all things, grey areas and overlap between the two groups). Tumor suppressor genes (*BRCA1* and *BRCA2* are the examples used) are discussed in the context of an inherited predisposition and contrasted with oncogenes. Gene fusions (including the Philadelphia chromosome in chronic myeloid leukaemia) are used as examples of acquired changes. Some of the principal diagnostic methods employed for genetic testing in cancer are described briefly, including methods for chromosomal analysis (karyotyping and FISH) that continue to play a vital role in this area. Genetic testing is an essential tool in establishing a diagnosis, helping to predict outcomes and directing treatment for cancers, and the utility of this testing in cancers is summarised.

Cancer as a Genetic Disease

There are many different forms of cancer, all of which share the essential property of being genetic in origin. Growth and development depend on cell divisions happening in a controlled way, and pruning back unwanted or damaged cells is as important to proper development as cell divisions happening in the right place, at the right time, and to the right extent.

Tissue growth is an immensely complex process, and errors in any of the genes involved in the processes controlling cell division, cell maturation or programmed cell death (apoptosis) could potentially lead to cancer—cellular proliferation out of control. Cancer is the result of an escape from the growth constraints that normally apply. Because damage to DNA accumulates throughout life, cancer becomes more likely with age, and the process can be accelerated by a hereditary predisposition or by exposure to environmental factors that increase DNA damage.

Tumors start from a single cell with relevant acquired or inherited damage; cancers are clonal, subsequent expansion deriving from that original cell. Further genetic changes are involved in the evolution of a tumor. These changes might increase growth of a cancer by directly promoting faster division, allowing escape from growth suppressors, avoiding apoptosis, evading the immune response or overriding normal limits on the possible total number of cell replications for a given cell. Changes may also promote uncontrolled growth by improving the environment for the tumor, for example by facilitating improvements to the blood supply, or by allowing the tumor to spread to other parts of the body. Effectively individual cells are subject to natural selection and evolution in an analogous manner to whole organisms, so an acquired change that allows a tumor cell to survive chemotherapy will give a selective advantage that will be inherited by any daughter cells. Although personifying cancer as an enemy may be successful as a strategy to raise funds to combat the disorder, in reality these are all blind processes, and cancer an inevitable consequence of our basic biology.[1] Any cancer has only developed as the result of the accumulation of a series of genetic changes in a cell, inherited by the daughter cells descending from that precursor.

The changes that allow cancer cells to gain a selective advantage are known as driver mutations, and there may be only a small number of these (perhaps 10 or fewer) needed to convert a normal cell into a cancer cell. Because of the growing loss of control over the genetic integrity of the cell, other changes will accumulate, some of which will be neutral in effect and can be considered as passenger mutations, irrelevant to tumor progression.[2] Driver mutations may

---

[1] Framing the experience of patients as that of someone fighting an opponent risks giving the impression that those who don't survive didn't struggle enough, when the outcome is dependent to a far greater extent on the underlying genetic cause and the availability of treatment rather than the attitude of the affected individual.

[2] Mistakes that occur during the replication of DNA are an important source of mutations. Given that each time a cell divides the entire genome has to be copied there is a huge potential for inserting an incorrect base into the newly copied sequence. An extremely effective proofreading machinery exists that keeps the error rate very low (although not zero—mutations will accumulate throughout life via internal events as well as by damage from external agents). Where the wrong base is inserted into a growing DNA strand or where other damage occurs it is usually possible to excise the incorrect

differ for different forms of cancer, but within a tumor type will tend to be the same in most patients, whereas passenger mutations will be more random. The fact that a small number of changes are crucial, and increasingly that interventions targeted against specific driver mutations are becoming available, is of central importance in designing rational genetic testing strategies. Diagnostic genetic testing is important in finding inherited mutations that can cause a predisposition to developing cancer. The approaches used for this scenario will be broadly similar to those used for testing other single gene disorders. However, it should be stressed that the majority of cancers are not due to an inherited predisposition, but are described as sporadic, originating from acquired DNA damage in what was previously a normal cell.

Whether inherited or acquired, driver mutations can take many different forms at the DNA level, from single base changes up to large-scale chromosome rearrangements (although large-scale changes would usually be acquired rather than inherited). Any change that abolishes the function of a gene involved in the control of cell division or programmed cell death, or that increases activity for a gene involved in speeding up cell division, may have the potential to facilitate cancer. Oncogenic (cancer promoting) events are not limited to changes directly altering the DNA; epigenetic changes to genes involved in growth promotion or growth regulation can be part of the accumulation of changes leading to cancer in just the same way as mutational events affecting the primary DNA sequence of genes. There tends to be global low levels of methylation (hypomethylation) in tumors, reflecting the loss of regulation for genes that promote growth (recall that high levels of methylation in the promoter regions for genes is associated with low gene expression; hypomethylation is associated with higher gene expression). The overall low level of methylation of the genome may be coupled with hypermethylation of specific parts of the genome that correspond to the promoter regions of genes involved in tumor suppression, which can be switched off by this mechanism. Testing methylation status for specific genes known to be involved in cancer progression is an increasingly important part of the genetic testing repertoire.

---

base or damaged region. One of the hallmarks of cancers is the early loss of function for genes with products involved in this proofreading process, which allows the accumulation of more damage, and the rapid acquisition of mutations that give cells a growth advantage. Genetic tests that detect evidence of this characteristic genomic instability are a useful adjunct to tests that directly detect driver mutations.

Inherited Cancer Predisposition

Inherited cancer predispositions, although devastating for the individuals and families affected, only account for a minority of cancers. It is difficult to quote an exact figure, and possibly slightly arbitrary to try to do so. (Even where a predisposition exists, acquired events are necessary. Conversely, it is unlikely that for most cancers generally labelled as acquired, the genetic background has no effect.) The difference between fully acquired and inherited cancers lies in the fact that the initial driver event that begins the process has already happened with an inherited cancer: all cells already carry the mutation, which effectively means there is an enormous pool of cells that have already taken the first step towards becoming a cancer cell. The subsequent changes still need to be acquired as for any cancer. The characteristics of inherited versus acquired cancers tend to be earlier onset and an increased risk of multiple independent tumors (reflecting the fact that all cells are already partly primed for this path) as well as the consistent involvement of cancers in the same tissues for a given inherited predisposition.[3]

It is important to note that in the majority of inherited cancers, the life-time risk of cancer may be very high, and elevated significantly above the population risk, but is rarely 100%. This reflects the fact that the inherited (germline) mutation is not enough in itself to cause cancer; successive somatic mutations are also needed, and this will depend on environmental as well as other genetic factors. The inherited mutation increases the likelihood that one or more cells will progress to cancer, without necessarily making this certain.

Inherited cancer syndromes can be divided into two broad classes. Uncontrolled cell division can be facilitated either by getting rid of control mechanisms (analogous to disabling the brakes in a car) or by directly promoting growth (analogous to stepping on the accelerator). In practice evolution of a tumor will involve both types of process. Those genes where loss of the gene product damages one of the mechanisms preventing uncontrolled cell growth—loss of braking—are classed as tumor suppressor genes. Those genes that when mutated help to allow unrestricted growth—acceleration—are classed as oncogenes (genes that promote cancer; properly speaking the normal unmutated version of the gene is a proto-oncogene, although oncogene tends to be used as a general term).

---

[3] High levels of exposure to environmental agents that cause DNA damage such as UV radiation, tobacco smoke or asbestos have a similar effect in increasing the likelihood of specific cancers, in this case by causing accelerated DNA damage in the exposed tissues, and making the accumulation of driver mutations more likely.

Tumor Suppressor Genes

There are some exceptions, but most hereditary cancer syndromes caused by mutations in tumor suppressor genes show a dominant pattern of inheritance, in that inheritance (or a *de novo* occurrence at an early stage of development) of a single mutation may be sufficient to confer a high lifetime cancer risk. However, as mentioned above, incomplete penetrance for these syndromes is the general rule, largely because further acquired changes are needed for cancer to develop. In particular, it has long been understood that although inheritance may show an apparent dominant pattern, at the level of the cell these mutations are recessive—there has to be a second event that knocks out the functional copy of the gene in question for tumor progression to start.[4]

Tumor suppressor genes are responsible for a small subset (in the region of 5–10%) of common cancers such as breast cancer and colorectal cancer, and the development of sporadic (as opposed to inherited) cancers of the same type often involves acquired mutations in the same tumor suppressor genes at an early stage. The involvement of the same genes in both inherited cancer syndromes and acquired cancers suggests a central role for the normal products of these genes in preventing uncontrolled cell growth in the tissues which are characteristically affected. This isn't always the case, but there are a number of well-established examples (including colorectal cancers), where acquired mutations in the genes known to be responsible for inherited forms of colorectal cancer turned out to be crucial early events in sporadic cancers in the same tissue.

BRCA1 and *BRCA2* as Tumor Suppressor Genes

The relatively well-known *BRCA1* and *BRCA2* genes (found on chromosomes 17 and 13, respectively) demonstrate some of the characteristic features of tumor suppressors. Mutations in these genes are relatively common; figures vary in different populations and in different studies, but mutations appear to be present in around 1 in 400 to 1 in 500 individuals in most populations. The gene products play an essential role in DNA repair and loss of this function therefore facilitates the accumulation of further mutations. Inheriting a single copy of an inactivating mutation in *BRCA1* or *BRCA2* is sufficient

---

[4] This is encapsulated in the two hit hypothesis (or the Knudson hypothesis after the originator of the theory, developed from Alfred Knudson's work on a tumor suppressor gene). A characteristic finding in tumors in these syndromes (where a single mutation has been inherited) is loss of the functional allele through a second event. This second event could be a single base change, but not infrequently is a large-scale change involving the loss of part of the chromosome in question.

to give a high lifetime risk, so the pattern of inheritance is autosomal dominant. Because a second event is needed to inactivate the normal copy of the gene, penetrance is incomplete. Mutations in these genes confer a high risk of cancers in specific tissues, most importantly breast and ovarian cancer in female heterozygotes. Risks for some other cancers, including prostate cancer in males, are also raised.

The population lifetime risk for women is 12% for breast cancer and 2% for ovarian cancer (these risks reflect figures for largely wealthy Western nations, and may differ for different populations). For women who inherit a *BRCA1* mutation the lifetime risk figures are approximately 70% for developing breast cancer, and 45% for ovarian cancer. For *BRCA2* the corresponding figures are 70 and 20%. It is estimated that in combination *BRCA1* and *BRCA2* account for around 3% of all breast cancers and 10% of all ovarian cancers—a huge burden. There are many different mutations in *BRCA1* and *BRCA2* which are known to predispose to hereditary breast and ovarian cancer; these are loss of function mutations, and there are multiple ways in which the gene function can be abolished. Testing for mutations in these genes is increasingly based on DNA sequencing using next-generation sequencing methods, often as part of a panel of genes. (There are other single genes which when mutated convey a high risk of breast cancer, but account for a much smaller proportion of cases than *BRCA1* and *BRCA2*.) In families where a pathogenic change in the DNA sequence has been found, subsequent familial testing (predictive tests for those unaffected or confirmatory tests for those with cancer) can be carried out using Sanger sequencing.

There are some relatively common mutations in *BRCA1* and *BRCA2*, either recurrent because the local sequence environment is conducive to errors (such as in short repeats within the sequence, or a run of the same base) or because of historical founder effects.[5] In populations with a strong founder effect, testing for specific mutations may be a useful initial strategy, but overall since the mutations tend to be distributed around the whole coding and regulatory regions of the genes, full sequencing of both of the genes makes sense. This does have the downside that all variants will be detected, and in many cases it will be uncertain as to whether these are causative (most patients tested will not have a pathogenic *BRCA1* or *BRCA2* mutation). It is essential that these changes are interpreted with caution. Patients may choose to have an unnecessary prophylactic mastectomy or oophorectomy based on the

---

[5] Direct-to-consumer tests are available for some of these relatively common *BRCA1* and *BRCA2* changes. Where a mutation is found, getting this sort of information almost out of the blue and without any genetic counselling is likely to be a huge shock for the individual. There is also a knock-on effect for the healthcare systems that will be dealing with the downstream consequences.

results of a misinterpreted genetic test. Conversely, patients may miss out on potentially life-saving screening if they are told they haven't inherited the familial predisposition, when in fact they have been tested for an irrelevant variant. Where the status of a change in the DNA is uncertain it is imperative that this is clearly stated in any diagnostic report (Chap. 7 includes more detail about the lines of evidence used in assigning pathogenicity). The state of knowledge is better for *BRCA1* and *BRCA2* mutations (although by no means complete) than for some rarer conditions, because of the huge numbers of tests carried out, international sharing of data, and systematic functional studies on the effect of specific changes.

Oncogenes

In contrast with tumor suppressor genes, where inactivating mutations compromise the control of cell proliferation, changes in oncogenes increase the activity of gene products that promote cell division. There are also relatively few instances of inherited conditions due to mutations in oncogenes, although there are some important examples of genes where inherited activating mutations are found. Most of the oncogene mutations described below occur as acquired changes.

Activation of oncogenes can occur in a number of ways. Single base changes can alter the conformation of the resulting protein, or otherwise change the way that the gene product responds to growth signals so that effectively the gene is always switched on. As gain of function mutations such changes tend to be very specific; whether these are acquired or inherited, it is often possible to design rapid targeted tests to detect the changes, as opposed to the sequencing of whole genes often needed for tumor suppressors.

Another possibility is gene amplification: growth is promoted by brute force as multiple copies of a gene can produce enough of the associated protein to overwhelm the normal control mechanisms. There can be hundreds of copies of an oncogene in a cancer cell, amplified to the extent that these are visible as physical changes to chromosomes and easily detectable by methods such as MLPA (see Chap. 4), array analysis, or FISH (see below).

Oncogenes can also be activated by large-scale chromosomal rearrangements. Reciprocal exchanges of material between chromosomes (called translocations) can create gene fusions, novel hybrid genes where the regulatory regions of one gene are fused to the business end of another gene, decoupling this second gene from the normal regulatory controls and therefore potentially allowing unrestricted growth. Any rearrangement that moves an oncogene from its normal chromosomal position to an environment where the local genes are expressed at a high level can also increase oncogene expression, and as mentioned earlier hypomethylation can effectively upregulate oncogenes.

Cytogenetics and Cancer Testing

Most of the testing described in the book so far has been at the DNA level, and this is perhaps now the main emphasis for genetic testing. However, there are specific advantages to analysis at the level of the chromosome in a number of important scenarios, and particularly for acquired cancer, which as a cause of morbidity and mortality far outweighs that caused by single gene inherited conditions.

Chromosome Analysis

Standard chromosome analysis (karyotyping) involves cell culture and light microscopy. Nucleated blood cells from patients (not the red blood cells, which contain no nucleus and therefore no chromosomes) can be artificially grown in the laboratory, cultured in a way that makes chromosomes maximally visible under a light microscope. The same applies, with appropriate changes to the culture methods, to cells from tumors or other tissues (although chromosome preparations from cancer cells are often of poorer quality that those derived from normal cells). Staining of the chromosomal material in the cultured cells is used to produce a banding pattern based on the chromosome structure that makes each chromosome uniquely identifiable to the skilled eye. Photographic or more usually now digital images of the chromosomes can then be sorted into their appropriate pairs, plus the X and Y chromosomes. It therefore becomes possible to detect changes in chromosome number (aneuploidies), gains or losses of parts of chromosomes, inversions of a section of a chromosome, or translocations (see Fig. 5.1).

The value of chromosome analysis is limited by the resolution of the technique—in terms of bases of DNA, only very large changes can be detected (although this does have a positive consequence, in that the majority of changes found are likely to be significant, involving many genes). However, such large changes are recurrent findings in cancer, and specific rearrangements or changes in the number of copies of chromosomes detectable by

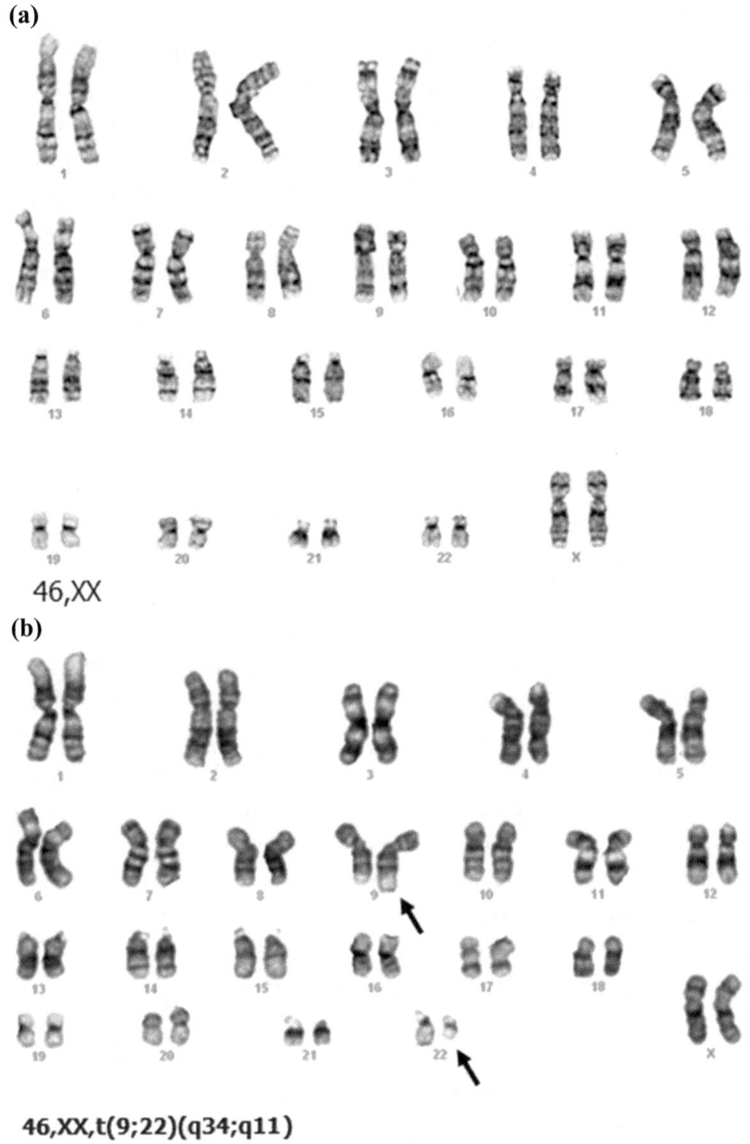

(a)

46,XX

(b)

46,XX,t(9;22)(q34;q11)

Fig. 5.1 Part **a** shows a normal female chromosomal complement (described as 46, XX). Part **b** shows the chromosomes from a cell in a female patient with CML and the *BCR-ABL* fusion (see section below; the Philadelphia chromosome, arrowed at the bottom of the figure, is the small derived chromosome paired with the normal chromosome 22, while the reciprocal product, also arrowed, is paired with chromosome 9). The nomenclature at the bottom left of the picture describes the chromosomes involved in the translocation and the positions of the breakpoints in each. Photographs courtesy of Dr Nick Bown, Northern Genetics Service

karyotyping can be vital in forming a diagnosis and giving an idea of prognosis. The great advantage of this type of analysis is that a karyotype gives positional information for deletions, duplications and translocations which is not available from most standard DNA methods (although advances in sequencing methods are beginning to address this). For large enough rearrangements, you can see exactly where the change is located on a chromosome or chromosomes directly, rather than (for example) detecting an extra copy of a gene by MLPA or array analysis which tells you nothing of the location of that extra copy.

The Philadelphia Chromosome

Many acquired translocations seen in individual types of cancer are known to be specific to those cancers, and are found recurrently. A well-known example of a chromosomal rearrangement driving cancer by the formation of a gene fusion was originally detected in a lab in Philadelphia, hence the Philadelphia chromosome. The acquired change is found in patients with chronic myeloid leukaemia (CML) and involves the reciprocal exchange of material between chromosomes 9 and 22 (the translocation being described as t(9;22); see Fig. 5.1).[6] Breaks in chromosomes 9 and 22 are followed by the creation of two hybrid chromosomes, the two parts of chromosome 9 joining to the two parts of chromosome 22, rather than re-joining at the original break points. (Such events are of course rare, but if the cell gains a proliferative advantage then the error will spread, and cancers start from a single cell.) The specific breakpoints involved lead to a gene fusion between the *ABL* oncogene from chromosome 9 and the *BCR* gene from chromosome 22. *ABL* is one of a class of genes with protein products known as tyrosine kinases that play a pivotal role in cell differentiation and in cell division. Linking the coding part of the *ABL* gene with the regulatory region of the *BCR* gene means that *ABL* expression is no longer controlled by the appropriate signals, and cancer is the result. Activation of tyrosine kinase genes is a common driver event in a range of cancers, and one of the biggest recent advances in cancer treatment has been the development of tyrosine kinase inhibitors, small molecules that specifically interact with the active site in tyrosine kinases and block their effect. The drug Imatinib is one of the first examples of a successful cancer treatment based on a targeted approach, rather than the sledgehammer of general chemotherapy or radiotherapy.

---

[6] There are several different classes of leukaemia. Chronic, as opposed to acute, describes leukaemias with a slow disease progression, and myeloid refers to the particular lineage of the blood cells involved.

Rapid Detection of Specific Gene Fusions and Other Chromosomal Rearrangements in Cancers by FISH

A cytogenetic method of huge value that combines positional information and better resolution than standard chromosome analysis is fluorescence in situ hybridization (FISH). In this technique fluorescently labelled DNA probes for a region of interest are allowed to find their complementary sequences in cell preparations from cancers or from other tissue. (The cells do not have to be grown in cultures, giving an immediate advantage in terms of speed of results when compared to karyotyping, which is much more laborious—rapid diagnosis is often essential in cancer treatment.) These probes will hybridize to the chromosomes in the cells and crucially can therefore give positional information about large-scale DNA rearrangements affecting that chromosome, using specialised microscopy to detect the fluorescence. If two probes from different genes, which are normally found on different chromosomes, are seen to localise together in all cancer cells this is likely to be due to a fusion of these genes. Alternatively, if two probes that normally co-localise are split apart this is equally indicative of a chromosomal rearrangement. Such fusions and other rearrangements are characteristic of many cancers, and detection of these changes is essential for diagnosis, prognosis, monitoring cancer progression and (increasingly) directing treatment as mentioned for Imatinib above. FISH can also be used to detect activation of an oncogene by amplification, as the fluorescent signal will increase in proportion to the number of gene copies.

Figure 5.2 shows the results for FISH analysis for a gene fusion. The probes for the genes involved each come in two parts, both labelled in red for one gene, and both in green for the other. In a normal cell, the two red probes will hybridize to adjacent regions in one gene, and the two green probes to adjacent regions in the second gene, each on their respective chromosomes. As there are two copies of the relevant genes, in a normal cell four regions of colour would be expected, two green and two red. The assay works by designing the probes to flank points on both chromosomes where specific chromosome breaks have been seen as a recurrent feature in cancers (this type of assay is therefore used to confirm or exclude the presence of known rearrangements). In the example shown in Fig. 5.2, breaks in chromosome 15 and 17 are followed by the two parts of chromosome 15 joining to the two parts of chromosome 17 forming hybrid chromosomes, one of which carries an oncogenic gene fusion (the same basic mechanism as that leading to the *BCR-ABL* fusion, which can also be detected by FISH). The fusion can be

(a)

(b)

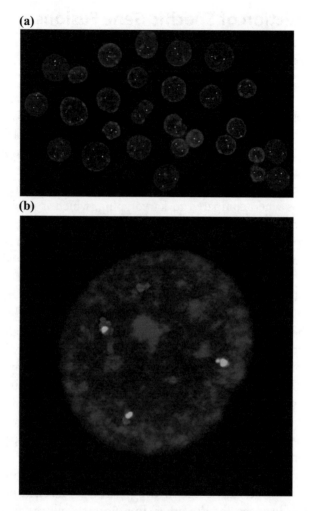

Fig. 5.2 Fluorescence in situ hybridization (FISH). Part **a** shows a mixture of cancer and normal cells from a patient with an acute leukaemia, and part **b** a single cancer cell at a higher level of magnification. Fluorescent probes for two different genes (*PML*, labelled in red, and *RARA*, labelled in green) have been added to the cell preparation. These two genes become fused in this specific type of leukaemia, and this fusion generates a hybrid gene, indicated by the co-localisation of the red and green probes, which is no longer properly regulated, and drives the cancer. The genes in this case are carried on chromosomes 15 and 17, and the gene fusion can occur when one chromosome 15 and one chromosome 17 undergo a break, with the breakpoints occurring within the *PML* gene on chromosome 15, and within the *RARA* gene on chromosome 17. Photographs courtesy of Gavin Cuthbert, Northern Genetics Service

confirmed by co-localisation of the red and green probes, as seen in part **b** of Fig. 5.2. The normal chromosomes 15 and 17 are indicated by separate red and green signals. Depending on where chromosomes 15 and 17 happen to be located in an individual cell, the red and green signals might be close together purely by chance (the relative position of the chromosomes within the nucleus can be considered as effectively random for this type of assay). It is therefore necessary to examine a number of cells to confirm that chance proximity is excluded, and establish the presence of a gene fusion. Direct FISH analysis is a powerful method to allow very rapid diagnosis for specific cancers.

Genetic Testing in Cancer Diagnosis and Treatment

Although only one of a heterogeneous range of investigations, spread across many medical disciplines, genetic testing has obvious value in diagnosing cancers, and knowledge of the specific changes involved can also provide powerful prognostic information. This can inform treatment in that if the prognosis is poor and disease progression likely to be rapid, more intensive treatment may be indicated (or palliative treatment more appropriate). Increasingly, genotyping of tumors is used in stratifying treatment, as specific drugs are targeted to specific mutations. Genetic testing is also critical for monitoring cancers, using highly sensitive (usually PCR-based) methods that can predict relapse by detecting very low levels of cancer cells. Monitoring might include testing for drug-resistance mutations (tumors constantly evolve, and genetic changes that allow resistance to treatment will have an obvious selective advantage). Where resistance mutations have been found second line treatments, if available, can be employed. The advent of next-generation sequencing has, among other improvements, facilitated the detection of very low levels of tumor DNA in blood (see Chap. 7), which might allow diagnosis without knowing where a tumor is located, obviate the need for invasive surgical sampling, and even allow screening before any clinical symptoms manifest. In terms of inherited cancers, a genetic diagnosis in an affected individual can allow predictive testing in at-risk relatives, with the options of intensified screening, preventative measures or prophylactic surgery for those who have inherited a mutation.

6

DNA Testing, Genetics and Identity

Summary In the first part of this chapter, the basic DNA testing method employed in establishing identity is described (using PCR amplification of variable short tandem repeats). Some of the applications that depend on testing these unique combinations of variable DNA sequences in an individual are introduced (including confirmation of sample identity, controlling for errors in prenatal diagnosis, establishing zygosity in twins, and monitoring bone marrow transplants). Uses of these variable repeats in areas other than identity testing are briefly mentioned (aneuploidy testing and linkage analysis). DNA testing is shown to have obvious practical value in determining identity in the narrow sense of being able to discriminate between biological samples from two different individuals. The final part of the chapter explores broader questions of personal identity, and the role (if any) of genetics in illuminating concepts of identity at the individual, group and human species levels.

Identity Testing in the Diagnostic Genetic Laboratory

A significant part of the human genome consists of repetitive DNA, where the same sequence motif is present a varying number of times. These repeated regions can vary hugely in size, but the repeat types of main practical use in diagnostic genetics are known as short tandem repeats (STRs) which cumulatively account for approximately 3% of the genome. The STRs employed for diagnostic work are generally simple sequence repeats of between one and five bases. This definition would include the CAG repeat seen expanded in HD (and the repeats in other trinucleotide expansion disorders), but in most cases the STRs do not form part of the coding sequence or immediate flanking regions of a gene, and in general if these repeats have a function it is not as yet understood. Given the apparent lack of function, and the ease with which errors can occur when replicating simple repeat sequences, these regions will tend to vary between individuals in terms of the number of repeats at a particular point in the genome. Genotyping these repeats can therefore potentially give a genetic fingerprint unique to the individual. Any STR where the repeat alleles differ between two patient DNA samples can be used as an informative marker. STRs are widely distributed throughout the genome, and as outlined below provide the most practical laboratory tool for establishing identity.

STRs can be typed using essentially exactly the same method as that described for HD in Chap. 2: PCR amplification of the region including the repeat. As for HD, the specific primers used will be designed to be complementary to the unique sequence flanking the repeat region, and the size of the fluorescently labelled amplified DNA fragments, which depends on the number of repeats present, is assessed using capillary electrophoresis. There are various commercial kits available for identity testing, and one popular version uses a mixture of 15 different STRs spread across the genome, plus a marker that gives different fragments for the X and Y chromosomes and therefore establishes the genetic sex. Using a 15 STR kit, identity can be excluded or confirmed with a very high degree of probability. By chance, two unrelated individuals may share the same alleles at several of the markers tested, but statistically it is hugely unlikely that any two individuals taken at random will have the same genotype at all of the markers tested. In practice, any two individuals, apart from identical twins, can be expected to give different results with this method.

For legal work (in criminal cases or for establishing family relationships) it is usually necessary to calculate the odds against a chance match for two unrelated individuals using population figures for the frequency of the different STR alleles at each marker. For most medical applications, the calculated probability may not be required in the same way since the odds against a chance match are high enough under any plausible circumstances to avoid any meaningful doubts as to the confirmation or exclusion of identity. (For this type of test, the probability of a chance match is vanishingly low. Once the probability of a mistake due to a chance match is sufficiently low, this can be considered negligible in comparison to other possible sources of error.)

Diagnostic Applications for Genetic Identity Testing

Although the basic test remains the same, there are applications for identity testing in a range of different scenarios in a medical setting. The next part of this chapter lists some of these applications for STR analysis in a diagnostic laboratory. Identity testing is of course also of central importance to forensic work. Aside from the huge technical challenges relating to the need to test vanishingly small and often compromised samples (not usually applicable to diagnostic samples, although there are exceptions), forensic work uses essentially the same STR typing technology as the medical applications.

Direct Testing of Identity

Perhaps the most obvious application for identity testing is in a direct comparison between two samples to establish whether or not they originate from the same individual. Diagnostic laboratories may receive thousands or tens of thousands of samples per week, and occasionally circumstances arise where a patient sample has potentially been mixed up with another, or has been contaminated with a second sample. Where this is suspected, a known sample from the patient can be tested and the results compared with those from the possibly misattributed sample. Any error in sample identity can be confirmed or excluded. If there is a suspicion of contamination of a sample, the origin of the doubtful material can be shown definitively.

Family Relationships

Identity testing is of obvious use in establishing family relationships, paternity being the most frequent relationship in question. Although diagnostic laboratories may not be involved in legal paternity tests, it will occasionally be relevant to establish paternity (or other familial relationships) for diagnostic reasons. Paternity can be excluded where there is clear non-inheritance of alleles by a putative child, ideally for more than one marker (a single example of a discrepant STR might leave some room for doubt, given that these markers are subject to rare changes in repeat length on transmission). If the putative child shares at least one allele at each of the markers tested this is consistent with paternity, and the odds against a chance match would be high. More distant relationships are proportionately more complex to disentangle, but there are statistical methods available that could give the relative probabilities of (for example) two individuals being cousins as opposed to being unrelated. Great care is needed in this sort of test to allow for specific population effects (for chance historical reasons, some alleles may be more common in specific populations). Other testing is also available, such as looking at variable Y chromosome markers when investigating male-to-male descent.[1]

Identity in Twins

In most cases, a medical laboratory would not test whether or not same-sex twins were identical, as there would not be any clinical significance. Suppose, however, the situation arose where twins of the same sex are discordant for a congenital condition—one twin is affected with a severe disorder at birth and the other is unaffected. It is suspected that the condition is genetic. If the twins were fraternal or non-identical (the results of two different eggs being fertilised by two different sperm, hence dizygotic twins) they would be genetically no more alike than any other siblings, and the possibility of a genetic cause for the symptoms seen in the affected twin would remain open. (Non-identical twins will have inherited a different combination of parental genes, and also a *de novo* mutation could have occurred in the affected twin, either of which could explain the difference in phenotype.) If on the other hand the twins were identical (monozygotic, the result of a single fertilised egg dividing into two embryos) this might make a genetic cause much less likely, as the twins would share the same genome and therefore would be expected to both have symptoms if the cause were genetic (assuming that the condition was fully penetrant)[2] Zygosity testing would therefore be appropriate to inform any decisions as to whether further genetic testing would be helpful.

Monozygosity could be easily excluded, if one or more markers were found to be different on testing both twins. Confirming monozygosity is slightly less straightforward; clearly, children of the same parents are more likely to share the same marker alleles than would be the case for unrelated individuals. The odds against a chance match of all 15 STRs tested (still enormously high)

---

[1] In societies where the children take their father's surname, family names should track with the Y chromosome in males, with potential benefits for ancestry research, although this would be confounded either by less patriarchal habits or simply by non-paternity. Using the Y chromosome to trace male-to-male ancestry is a useful adjunct to the use of the mitochondrial genome to trace female ancestry as mentioned in Chap. 1.

[2] In reality even at birth monozygotic twins are not completely identical, due to differences in their local uterine environments and perhaps unequal partitioning of resources when the fertilised egg splits into two embryos.

could be calculated more precisely using Bayesian methods if parental samples were available to allow genotyping for the markers used.

Avoiding Errors in Prenatal Diagnosis

The potential consequences of an error in prenatal diagnosis are profound. If a prenatal test is being carried out for a severe genetic disorder, with the option of termination of an affected pregnancy, a false positive result could lead to an unaffected pregnancy being terminated, and a false negative result to the birth of a child affected with that condition. Prenatal diagnosis for genetic conditions largely relies on two main procedures to sample fetal tissue: chorionic villus sampling (CVS) and amniocentesis. CVS involves taking a small amount of placental material (chorionic villi form part of the placenta, which is a fetal rather than a maternal tissue) and extracting DNA or preparing chromosomes for analysis. Amniocentesis is a procedure that allows collection of fetal cells from the amniotic fluid, again giving access to fetal DNA or chromosomes. Both of these procedures are invasive, so as well as the direct implications for the mother and a small elevation in the likelihood of miscarriage, in both cases it is almost certain that at least a small amount of maternal tissue will also be collected.[3] There is therefore a risk that the analysis will give the maternal genotype rather than the fetal genotype, which could cause either a false negative or false positive result depending on the pattern of inheritance in that family.

Because of this risk, a maternal cell contamination (MCC) check will always be carried out. This involves testing maternal DNA for a series of STR markers alongside the DNA extracted from the CVS or amniocentesis, and looking for evidence of MCC. One of the maternal alleles will have been inherited by the fetus, along with the paternal allele for that particular STR. If the fetal material is contaminated to any significant degree, both maternal alleles will be present for every STR tested, and where those markers are informative this can easily be detected (see Fig. 6.1). This method also allows quantification of the amount of contamination, and depending on the analysis planned, above a certain level MCC might make a test of a prenatal sample unsafe. Despite the associated physical and psychological problems, a new sample would have to be taken. This is clearly not ideal, especially given the extra burden associated with a later termination of a pregnancy, but may be unavoidable. In some cases, the results of the direct genetic test carried out

---

[3] See chapter 7 for the use of next-generation sequencing in non-invasive prenatal diagnosis.

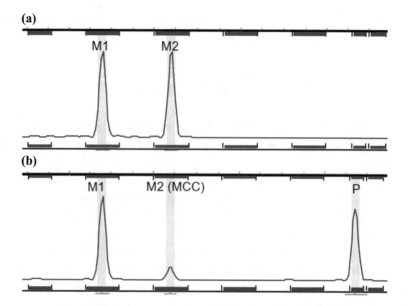

Fig. 6.1 Part **a** shows the results for a single STR in a maternal DNA sample. The two maternal alleles are labelled M1 and M2. Part **b** shows the results of testing DNA from a fetal sample. The paternal allele is labelled P. In addition to the M1 allele, which has been inherited by the fetus in this pregnancy, the M2 allele is also present, at a lower level, due to contamination with maternal tissue. Depending on the nature of the prenatal test, this degree of contamination may mean that giving a result for this sample is not safe. (The M1 allele is present at a higher level than the paternal allele, reflecting the fact that this peak incorporates the contribution of both the fetal DNA and the contaminating maternal DNA)

on the fetal DNA will themselves exclude significant MCC (for example if a paternal mutation is detected at the appropriate level, or in any other instance where the fetus has a clearly different genotype from the mother).

Monitoring Bone Marrow Transplants

A bone marrow transplant (BMT) can be a life-saving treatment for patients with cancers such as leukaemia or lymphoma, or individuals with severe immune disorders. One of the ways to monitor the success or otherwise of a BMT is to test DNA from the recipient (pre-transplant) and the donor, and look for STR markers that differ in the two. It is then possible to test post-transplant samples to see whether the bone marrow cells (or any blood cells made by the bone marrow) are 100% donor if the transplant has been successful, 100% recipient if rejected, or a mixture of both genotypes (technically called chimerism, describing the simultaneous presence of cells from two different individuals). Monitoring changes in these proportions over time can be important in helping to decide on whether further clinical interventions are needed (see Fig. 6.2). For example, an increase in recipient cells in a cancer patient may give warning of a relapse after treatment. Patients may be followed up for years using this type of testing (in contrast to the other types of identity testing described, where the genotype would only need to be tested once).

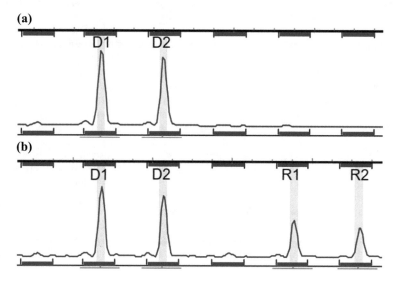

Fig. 6.2   Part **a** shows the result of testing a post-transplant sample for an informative STR where only donor alleles (D1 and D2) are present. Part **b** shows a subsequent sample where recipient alleles (R1 and R2) are also present. Testing a series of samples over time can establish whether there is a trend towards loss of the genotype corresponding to the donor cells, and if so this trend can be quantified

Box 6.1 Other Uses of STR Testing

In addition to the use of STR markers for the types of assay described in this chapter, which all involve aspects of identity testing, this versatile tool can be used in other ways in the diagnostic genetics laboratory. Two of these are briefly described below.

Testing for the gain or loss of whole chromosomes (aneuploidy testing) can be rapidly carried out using STR analysis. This is frequently used in testing prenatal samples for the common trisomies (trisomy 13, 18 and 21). Several STRs are tested for each of chromosomes 13, 18 and 21. The presence of three rather than the usual two chromosomes in one of these trisomies would be indicated by finding three detectable peaks for an STR, or by detecting two peaks with one present in a 2:1 ratio compared to the other. For any one marker it is always possible that all three chromosomes happen to have an allele with the same repeat length—the marker is uninformative—but if enough markers are used it is unlikely that all will be compromised in this way. Similarly this technique can be used to detect the more common but usually less significant sex chromosome aneuploidies (any combination other than XX or XY, such as XXX, XYY or X0) using a selection of markers from the X and Y chromosomes.

STRs can also be used for following single gene disorders through a family. The use of linkage analysis was at one time a staple in diagnostic genetics, but advances in testing, and particularly in sequencing, have made this sort of approach almost (but not completely) redundant.[4] Ideally to do predictive or carrier testing in a family you would detect the causative mutation, and test relatives accordingly. If the clinical condition is clearly caused by a single known gene, but the mutation has not been found (an increasingly rare situation), it may be possible to identify informative STRs that are found very close to the gene involved. In addition to at least one eligible informative marker, this approach requires a suitable family structure to establish which STR allele happens to be closely associated with the high risk copy of the gene in affected individuals in that family. This would allow indirect predictive or carrier testing in the at-risk individual (and is also the basis for HD exclusion testing: see Box 2.1). Even where these requirements are met, linkage analysis suffers from a further disadvantage compared to direct testing for a mutation. This is due to the phenomenon of recombination. Chromosomes are not inherited as the same enormous length of DNA through the generations. In the process of meiosis (which reduces the two copies of each chromosome to one in the production of eggs and sperm) material is swapped between pairs of chromosomes. The maternal chromosome 7 that you inherited is a patchwork of the two chromosome 7s found in your mother. This has the effect of helping to generate variation and new gene combinations. In the context of linkage analysis it also means that a previous association between a particular marker allele and a gene carrying a mutation might be broken up between generations. The likelihood of this happening is proportional to the distance between

---

[4] Genetic linkage refers to the likelihood that DNA sequences that are physically close together on a chromosome tend to be inherited together (in the absence of recombination, see above).

the marker and the gene in question, and as well as using very close markers there are ways round this, for example using markers found on both sides of the gene, but the multiple problems involved mean that linkage analysis is very much second best to a direct test.

DNA as a Marker of Unique Personal Identity

The utility of DNA testing to establish identity in the diagnostic scenarios described above is evident. It is less clear whether there is any value in attempting to use genetics to clarify our understanding of ourselves as unique individuals, as members of families, and of how we identify ourselves as a part of larger groups within humanity.

Taking the long view, each of us only exists because of an unbroken series of conceptions or equivalent events going back through all of our ancestors for at least 3.5 billion years or so—the odds against any particular individual existing are incalculable. The huge majority of possible individuals of any sort never came into existence. If you can get over the sheer improbability of existing at all, the concept of personal identity seems straightforward. Identity is often one of our unexamined assumptions, not something that it feels necessary to question. However, like many background assumptions, this concept is more elusive than it first appears.

Exactly what aspects of continuity guarantee a stable personal identity? Physically we are all in a state of constant change. Even apart from the obvious differences as we age, most parts of the body are in a continual state of renewal; beyond doubt you are not composed of the same atoms as you were at birth. Memory is central to a sense of identity, but is notoriously fallible (there is nothing to check your memory against in most instances, and when any tests of the reliability of memory have been done the results show recollection to be startlingly inaccurate). If you forget an event, are you no longer the same person who experienced that event? Stability of character or other psychological measures don't seem any more promising as a mark of identity. None of us remains the same person in that sense over time. If you have an accident or a progressive disorder that affects an important physical or psychological characteristic are you still identical to the person you once were?

Because of the centrality and relative stability of our genomes, our genetic make-up might be thought of as a candidate for a marker of personal identity. However, although apparently promising, it is not obvious that possession of a unique genome is in itself sufficient to define an individual. Nor is it necessary. In the special case of identical twins, sharing (initially) identical genomes does not make you the same person.[5] Identity is more complex by far than can be captured by a sequence of different chemicals, even when it is a very long sequence. The genome (and in particular the way that genes are expressed) does not remain completely stable over a lifetime. Genomic sequence is changed randomly in cells by damage accumulated over the years, and specific changes at the DNA level also occur (for example genes involved in the immune response are physically rearranged in the generation of antibodies). Stable changes to gene expression are now known to be caused by epigenetic modifications in response to the environment (see Chap. 7). The functional genome you have now is not the same as the one you initially inherited, and DNA is not necessarily better than any other single criterion as a marker of personal identity. Identity could be thought of as a process of perpetual change with some degree of continuity rather than something absolutely fixed. This applies to genomes as well as to other physical or psychological criteria for personal identity, although arguably the genome has a greater degree of stability throughout life.

Widening Circles

As for many of the most interesting but ultimately unanswerable questions, ideas of personal identity have evolved and been reformulated in various ways without reaching anything like a definitive answer. A genetic viewpoint might have some value in clarifying the coherence or otherwise of some of these ideas, but the field can also raise new questions (see Box 6.2). One way of looking at how identity is defined at different levels is through the idea of widening circles (usually discussed in the context of ethical interactions, moving from the closest to the most distant relationships). This captures the suggestion that our concerns start with ourselves as an individual, but then move out to include first our family, then our community, our nation, humanity as a whole, and then perhaps all sentient beings. This seems to correspond well to increasing genetic differences, in which case a genetic perspective might appear fruitful.

---

[5] Unless identical twins had exactly the same environment in every respect for their entire lives, their genomes will also differ due to acquired changes and epigenetic modifications.

A genomic point of view is probably most meaningful at the levels of the individual and the immediate family, and then at the level of the human species. The genome is clearly uniquely important to the individual, and our biological family relationships are of obvious significance. Although there are always some ambiguities, a definition at the level of species, which is ultimately a genetic definition, also picks out an important aspect of reality. However, for the intermediate points between immediate family and the human species genetics is of less value in defining groups, given that nation states are historical and geographical accidents, and populations beyond small localities rarely genetically homogeneous (whatever that might mean) within a state. There is a human tendency to sort things into categories, which is often useful but can be taken too far.[6] This can lead to a kind of genetic essentialism, where the genes are thought in some way to define or determine the essence of humanity. The risk of this way of thinking is that different genetic groups are put into different races, which in the past (and for some people the present) implied a hierarchy, with the race of the person doing the categorising at the top of the hierarchy.

An apparently more benign version of genetic essentialism lies behind the popularity of DNA tests that aim to give people an idea of their geographical ancestry, and allow them to say (for example) that they are 27% Viking, or that they had ancestors from various locations around the globe. However, there are some important theoretical limitations to the utility of this kind of testing. Even if you ignore these limitations, does this actually tell you anything meaningful about your identity, having found that you carry some variable markers, of no particular significance, which happen to be more common in a particular part of the world?[7] There is no such thing as the genetic essence of a Viking, and although the motivation for having this kind of test is generally harmless, the kinds of assumptions behind it are less benign in other contexts.

---

[6] This tendency was criticised by Richard Dawkins in a different context as an example of "the tyranny of the discontinuous mind", describing an apparent need to draw a clear dividing line where no such thing is justified.

[7] Essentially the methods used rely on comparison of variable sequences in the genome of the customer with databases of corresponding sequences from populations around the world. This approach to ancestry testing will get more accurate with improved representation of genomes in the databases and more sophisticated analysis, but as things stand the reference databases are skewed to some geographical areas at the expense of others. Different providers use different databases and therefore give different answers. What you end up with is at best a probability that you had an ancestor from a particular geographical region.

> **Box 6.2 Genetics and the Non-Identity Problem**
>
> Each of us only exists because a particular egg was fertilised by a particular sperm at a particular moment. If that had not been the case we would have no identity (however defined). Some of the ethical questions raised by genetics are made more difficult because the choices affect a person who does not yet and may never exist.[8] Suppose you have a situation where the technique of pre-implantation genetic diagnosis (PGD) is used to determine whether an embryo has inherited a familial mutation.[9] In this particular family, the parents already have an affected child, they themselves are affected and their wish is to have a second affected child, as the nature of the condition in this family is such that in their view another affected individual would fit in better. Would the family (and the medical teams involved) be wrong to select an affected rather than an unaffected embryo? The nature of the problem becomes clearer with the realisation that for the affected potential child, the choice is not between existence as an affected individual and existence as an unaffected individual, but the choice between existence as an affected individual and non-existence. The question then becomes whether life could be worth living for that individual, and in this hypothetical case the family would clearly think so.

Identity as a Member of Humanity

A level of identity that seems relatively non-controversial is our shared identity as humans, examples of the species *Homo sapiens*, but what exactly is the cut off at the genetic level that defines a human?[10] It is often quoted

---

[8] This type of scenario is an example of what was described as a non-identity problem by the philosopher Derek Parfit.

[9] PGD involves removing a cell or cells from a very early embryo and testing for a known genetic risk. The aim is usually to only implant embryos which have not inherited the familial mutation using the well-established techniques of assisted reproduction. This is seen by some families as a more acceptable alternative to standard prenatal diagnosis, as it allows selection of an unaffected embryo while avoiding the risk of potentially terminating an affected pregnancy after an adverse prenatal result. As well as the technical and practical difficulties, pre-implantation testing does still raise its own ethical issues, for example in the moral and legal status of the embryos that are not implanted.

[10] The definition of human may not be as straightforward as is generally assumed. We representatives of *Homo sapiens* are the only extant members of the genus *Homo*, but there have been many others in the past, partly known through the fossil record (including studies of ancient DNA), and increasingly from genetic traces in current human genomes, discovered through whole genome sequencing (see Chap. 7). There will undoubtedly be other extinct groups that have never come to light (at least so far—as more human genomes are sequenced there will be further findings). As recently as a few tens of thousands of years ago other close relatives shared the planet, and we now know that modern humans interbred with at least two groups of extinct members of the *Homo* genus, the Neanderthals and the Denisovans. Most current European populations still have a few per cent of their DNA derived from Neanderthals. Interbreeding with fertile offspring is one possible definition of a species; hypothetically, if either Neanderthals or Denisovans still existed, would we treat them as a human?

that humans are around 98% identical to chimps in terms of DNA (the precise percentage varies, depending on which methods and assumptions are used in the comparison). Although this does tell us something about our shared recent evolutionary history, it is a crude measure. Although on a global comparison of chimp and human genomes, similarity comes out at somewhere around the 98–99% level, a more useful comparison would be with how much functional DNA we have in common with other organisms. Essentially we share almost all of our genes with chimps, and the obvious biological differences are in large part due to the way the genes are expressed and interact. Rather than just looking at overall high percentage similarities with other apes and more distant relatives, a comparison at a much higher level of resolution is needed to give meaningful information (since there are only four bases in DNA, even a comparison of a human sequence with an entirely random sequence of DNA would show a 25% match). Some gene sequences for proteins involved in basic metabolic processes would be functionally identical across a whole range of species. Eventually, comparison of human and great ape genomes may give more clues as to what genetic differences are important. However, these differences won't necessarily involve the coding part of the genes, but may be found in regions that regulate the temporal and spatial expression of those genes. A definitive list of what tweaks make us human rather than chimpanzee will not be available any time soon.

Given how badly different human groups have been treated, we can probably guess the answer to that one. This comes back to the point raised above: where do you draw the line between human and non-human? At what percentage DNA similarity? Any answer will be to some extent arbitrary, and it may well be the case that future generations will be appalled by the way we currently treat our closest genetic relatives.

7

Out of Sequence: Genome-Scale Testing

Summary After a brief discussion of the evolution of whole genome analysis in diagnostic genetic laboratories (encompassing chromosome analysis, microarrays and sequencing approaches), the basics of Sanger sequencing are reviewed. The agreed universal system for mutation nomenclature is outlined. The central part of the chapter covers current uses of genome-scale sequencing. These include general mutation screening for rare disease and cancer, trio analysis, non-invasive prenatal diagnosis, detection of cell-free tumor DNA, finding new disease associations, long-read sequencing and comparative genomics. The relative advantages of genome-scale sequencing and targeted testing are compared, each being useful for answering different clinical questions. The final part of the chapter discusses possible limitations to the value of genome-scale techniques for mainstream medicine. Confounding factors include the variability of the human genome (some principles for assigning pathogenicity to sequence variants are outlined), and the underlying biological complexity of common disorders. Probabilistic factors and the problem of false positives in screening tests are also considered.

Whole Genome Analyses

To a degree genetic testing has come full circle. Some of the earliest testing involved a whole genome test when examining the chromosome complement, and we are now at a point where whole genome sequencing (WGS) is becoming part of the routine repertoire. The difference is in the degree

© The Author(s), under exclusive license to Springer Nature
Switzerland AG 2022
D. Bourn, *Diagnostic Genetic Testing*,
https://doi.org/10.1007/978-3-030-85510-9_7

of resolution that can be achieved, now at the single base level but originally involving large sections of chromosomes, at best detecting changes covering some millions of bases. Medically relevant genetic changes were being detected using chromosome analysis decades before DNA-based testing was available. However, the range of genetic testing expanded hugely with the development of methods that look more directly at sections of DNA, and ultimately single bases of DNA, rather than whole chromosomes. Even before the advent of routine WGS, traditional chromosome analysis had to a significant extent been replaced by microarray analysis as a way to detect copy number variants (deletions and duplications).[1] As an alternative to directly visualising chromosomes, microarrays allow a higher level of resolution. This approach (described briefly in Chap. 1) involves hybridizing fluorescently labelled patient DNA to single-stranded probes forming an array on a solid support, and interpreting the hybridization patterns to detect loss or gain of regions across the whole genome. The adoption of whole genome sequencing then takes this further, with resolution at the level of single bases. After a review of Sanger sequencing, this chapter concentrates mainly on the recent advances in the techniques used for genome-scale DNA sequencing (principally exploring the applications, increases in capacity and remaining limitations of these methods rather than the technical details).

Sanger Sequencing

Sanger sequencing has been one of the mainstays of diagnostic genetics for some decades, and still plays an important role. The technique was developed in the 1970s by Frederick Sanger, and depends on the use of slightly different versions of the four nucleotides that make up DNA. These modified nucleotides lack a part of the chemical structure that is essential for the incorporation of the next nucleotide when a new DNA strand is being synthesised from a single-stranded template. The effect is that when one of these modified nucleotides is incorporated into a growing DNA strand, further elongation is blocked; the chain is terminated, but only where that particular modified nucleotide, A, G, C or T has been inserted. Using carefully controlled reactions, it is possible to generate from a target DNA sequence new DNA fragments that terminate at every possible position. If these can be separated by size using electrophoresis, a ladder of fragments can be assembled that will give the sequence complementary to the target DNA. Prior to the

---

[1] Although traditional chromosome analysis is now used less as a front-line test, the skill is still needed for some vital specialist applications, particularly in cancer genetics as described in Chap. 5.

introduction of PCR, this process was relatively complex, and early Sanger sequencing also needed radioactive labelling. Current methods use PCR to rapidly produce precise copies of a region of patient DNA, for example amplifying an exon of interest in a particular gene. These PCR products are then used as the template for sequencing (note that as the PCR amplifies both copies of a gene, effectively both copies will be sequenced simultaneously). Rather than radioactive labelling (which meant that a separate reaction had to be carried out for each of the four nucleotides, as all were labelled in the same way) the terminating nucleotides are labelled fluorescently, each with a different colour. In the sequencing figures in this book an A will always be shown as a green peak, C as blue, G as black and T as red. When coupled with capillary electrophoresis, fluorescent Sanger sequencing allowed the completion of the Human Genome Project. However, this huge collaborative effort took years to complete—Sanger sequencing is inherently limited in capacity.

Diagnostic Sanger Sequencing Applications

Where genetic conditions are always due to the same mutation, or where there are a number of common mutations, using targeted tests such as those described in Chaps. 2, 3 and 4 may be the most timely and economical strategy to adopt. There are numerous other methods that lend themselves to high throughput simple genotyping for known mutations. This includes the group of techniques that come under the heading of minisequencing. Assays can be designed to interrogate just one or a few bases at the site of a known mutation using Sanger (or equivalent methods) for rapid and cheap testing.

However, such methods are not applicable where genetic disorders are caused by many different changes (and in some cases more than one gene needs to be tested for a given disorder). Prior to maximising the efficiency of Sanger sequencing, a number of elegant methods were in use to pre-screen for possible unknown changes before final confirmation by sequencing. However, once fluorescent sequencing advanced beyond a certain point it became untenable to use these methods, and workflow is simplified and made more efficient by putting everything through the same process in bulk (and Sanger sequencing was always needed in any case as the final step in determining the exact nature of a base change found by a screening method). For some years Sanger sequencing was the method of choice for detecting mutations in genes where a range of different sequence changes lead to disease. However, although Sanger sequencing was heavily used for sequencing whole genes, or even small panels of genes, for all but the smallest genes multiple PCR fragments are needed to cover the whole of the coding sequence, and this becomes progressively more cumbersome as the number of fragments increases. Furthermore, each sequenced target has to be subjected to capillary electrophoresis, adding another step. Most large-scale gene sequencing is now carried out through next-generation sequencing, as detailed below. Although Sanger sequencing is used less and less for identifying unknown mutations, it is still hugely useful in confirming mutations found by other methods, and essential for testing family members for these mutations. A bespoke Sanger assay for any sequence change can be designed and carried out rapidly and cheaply.

It is often necessary for a diagnostic laboratory to test a patient for a sequence change originally detected in a relative of the patient by another laboratory, or found as part of a research or genome project. If you know exactly what the sequence change is, and where it is located, it is routine to design a PCR assay that will specifically amplify the region in question, and then use Sanger sequencing to confirm the presence or absence of the mutation. This only works if everyone is speaking the same language; there has to be a universally agreed, completely unambiguous system of mutation nomenclature, to allow safe diagnostic testing (see Box 7.1).[2] If you test for the wrong mutation, you almost certainly won't find anything, and you could be missing the actual mutation and therefore give a false negative result. Ideally, you would safeguard against this by using a positive control—a family member known to have the mutation that you would then detect. Such controls are not always available, but even when they are you still need the information about the mutation in a form that allows you to design the assay.

Figure 7.1 shows the results of this type of test, in this case using fluorescent Sanger sequencing for a predictive test in a patient with a family history of colorectal cancer. A heterozygous mutation has been found in an affected relative of this patient in the *MLH1* gene, which is one of several genes known to predispose to this condition. The sequencing results show that this patient is also heterozygous for the mutation and is therefore at risk of developing the disorder. As discussed, both alleles are sequenced simultaneously following PCR amplification of a single exon of the gene. Therefore where a single base change is present two peaks are overlaid on each other, reflecting the presence

---

[2] Legacy (non-standard) nomenclature is still occasionally used where practitioners have become familiar with the original ways of describing some common mutations, as in the example of delta F508 in cystic fibrosis (see Chap. 3). The best practice is to move entirely to the standard system, but some habits are proving hard to shift.

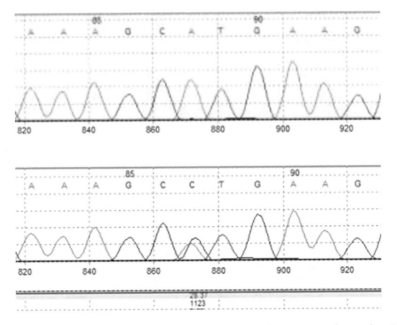

Fig. 7.1 The upper trace shows part of the normal sequence from the *MLH1* gene. The lower trace shows the results of Sanger sequencing in a patient with a heterozygous change in this gene, described as c.935A > C p.(His312Pro)

in this case of an A (the normal allele) and a C (mutated allele) at the same position in the two copies of this gene (more complex patterns are generated by small deletions or duplications; see Fig. 7.2). This mutation is described as c.935A > C p.(His312Pro) using the conventions set out in Box 7.1.

Box 7.1 Mutation Nomenclature

The universally agreed format for naming mutations is that implemented by the Human Genome Variation Society (HGVS, http://www.hgvs.org). This allows a precise description of a change at the DNA level (in the nuclear or mito-chondrial genomes), at the RNA level or at the protein level. The guidelines necessarily cover a huge range of scenarios, but the essential point is that a system exists that allows anyone to describe a mutation in a way that is completely clear to any other investigator. Figure 7.1 showed a patient with a single base change in the *MLH1* gene. This can be described using HGVS nomenclature as c.935A > C p.(His312Pro), using accession number NM_000249.3. This isn't as complicated as it looks. The accession number tells you which unique reference sequence has been used to assign this descrip-tion. The reference sequence is a universally available curated sequence that defines the normal sequence for this particular gene. As previously described the choice of a reference sequence tends to be a historical accident, in that it depends on which variants were carried in the first published sequence, but

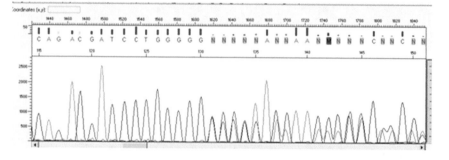

Fig. 7.2 Part of a Sanger sequencing trace showing the presence of a heterozygous deletion of one base. The final G in a run of 6 Gs has been deleted on one allele of this gene. This has the effect of shifting the subsequent sequence by one base compared to the normal allele. The mixed pattern of peaks beyond this point is generated by the simultaneous sequencing of two alleles that are one base out of step, and although the pattern looks complex it is usually straightforward to work out what the change is by a careful comparison with the normal sequence. Image courtesy of Dr Ciaron McAnulty, Northern Genetics Service

once chosen this is set in stone, so everyone is using the same comparison. Let's look at each of the other terms in turn. The prefix c tells you that this description is at the cDNA level.[3] Because the cDNA corresponds to the coding sequence of the gene, it makes a convenient reference point to use when describing mutations.[4] In this example c.935 denotes base number 935 in the cDNA sequence, and A > C that the base in the reference sequence at this position is normally an A, but in this case has been replaced by a C. This information allows you to predict that at the protein level this base change would cause the amino acid histidine at codon 312 to be replaced with the amino acid proline, His312Pro. The brackets in the description represent an acknowledgment that this is a prediction of the effect of a change, rather than a direct finding. The nomenclature includes formats flexible enough to describe not just single base changes in exons but also changes in introns, as well as deletions, duplications, and more complex rearrangements.

---

[3] The prefix g denotes genomic DNA, m mitochondrial DNA, r RNA and p the protein level. The same change could equally legitimately be described as g.37,020,360A > C at the level of genomic DNA (with the appropriate genomic reference sequence), but this formulation is perhaps less intuitively obvious.

[4] A DNA copy made from an mRNA molecule is referred to as complementary DNA or cDNA. Recall from Chap. 1 that in the process of transcription a messenger RNA (mRNA) copy is made corresponding to the exons of the gene being expressed. RNA is a less robust molecule than DNA, so it is often necessary when studying RNA in the laboratory to effectively reverse the transcription process using an enzyme called a reverse transcriptase to make the cDNA. The sequence of the cDNA is therefore the same as the sequence of the exons that coded for the mRNA.

Next-Generation Sequencing (NGS)

The brief introduction in Chap. 1 touched on the use of the term next-generation sequencing, which refers to a range of techniques that rely on massively parallel sequencing of short DNA fragments potentially spanning the whole genome, as opposed to the use of Sanger sequencing, one specific fragment at a time. Powerful software then allows assembly of the reads into contiguous sequences. The resulting enormous increase in sequencing capacity had an almost immediate impact on the practice of diagnostic genetics, and some of the applications (rather than the technical details of the methods) will be described below—this won't be an exhaustive list and undoubtedly new avenues are still to be explored. It is difficult to overstate the degree to which sequencing capacity has improved. The capacity for genome-scale analysis has moved much molecular biology laboratory activity away from doing piecemeal bench work and towards automated generation and analysis of huge data sets.

Diagnostic Applications of New Sequencing Technologies

Early applications of NGS were useful in replacing Sanger sequencing where the existing technology was stretched, for example in sequencing large genes or small panels of genes. (This kind of application employs a preliminary step to isolate the fraction of the genome that is of interest for the sequencing test in question.) As more powerful platforms became available, routine use was expanded to include large gene panels (covering many hundreds of genes), then whole exomes and ultimately whole genomes.[5] With the more sophisticated methods and analysis tools now available, whole genome sequencing also gives information about large deletions and duplications, beyond changes

---

[5] The term exome, analogous to genome, describes the sum total of all of the coding exons in human genes. The exome is very much smaller than the total genome (approximately 1–2% of the size). As most mutations are found in the exons and immediately flanking regions, whole exome sequencing (WES) seems an attractive approach in terms of cost and relative simplicity of data analysis. This technique has been applied with great success for rapid analyses in newborn children with a possible genetic disorder (see below). However, whole genome sequencing (WGS) may prove to be the more fertile strategy in the long term. Firstly, improvements in the technology and in the analysis software have reduced the cost of sequencing and the difficulty of the analysis. Secondly, WGS (in theory at least) should capture all mutations, including those found at some distance from the coding regions of the gene (rare, but they do exist, and the rarity in part may be because previously there was no way to look for them). Finally, WGS has proved to be technically better at covering regions of interest, including the exons.

at the single base level. The advantage for the patient with a possible rare inherited condition is that rather than having to wait perhaps years for sequencing of multiple candidate genes, all the genes can be sequenced together. Even if the result is negative or equivocal, the delay will be much shorter. As well as doing what Sanger sequencing does but more quickly, there are features of the new technology that allow for new approaches, and some specific examples are given below.

Trio Analysis and New Mutations

Severe congenital genetic disorders are often caused by new mutations. This is an inevitable pattern, since a life-limiting condition might make it impossible for patients to have children themselves. Prior to next-generation sequencing there was often very little that could be done to allow a precise diagnosis to be made, unless the symptoms were such that there was a likely candidate gene or genes. It is now routine in this scenario to carry out trio analysis, which involves whole exome or whole genome sequencing from an affected child and both parents. It is then possible to identify all of the *de novo* variants in that child by looking for any sequence changes not present in either parent. There will be multiple changes found, but the number is likely to be between about 50 and 100, and it is manageable (although not trivial) to examine these and see which if any disrupt a suitable candidate gene. WES in trios is already being used to great effect in diagnosing genetic disease in severely affected neonates. Similar successes have been reported in analysing fetal samples where abnormalities have been detected on ultrasound scans. Even if the diagnosis does not lead to treatment, it may well help the family and clinicians make decisions about the best options. If a mutation is found to be *de novo* the recurrence risk is low (but not zero—germline mosaicism is sometimes seen in one of the parents). As penetrance and expressivity vary for dominant conditions, it is not impossible that one of the parents may be heterozygous for the mutation causing symptoms in the child, but are themselves unaffected or mildly affected.

Trio analysis can equally well lead to the diagnosis of recessive conditions, and in the event that both parents are carriers of a mutation that is homozygous in the affected child, there are immediate implications for possible recurrence.

Genomic Analysis in Cancer

The increasing capacity for routine WGS allows the possibility of a full comparison of a tumor genome with the original genome from the affected patient (the sequence of the patient's original genome being ascertained from any suitable non-tumor tissue). All of the genetic changes that have occurred in the evolution of that tumor can in principle be captured. As mentioned in Chap. 5, not all of the genetic changes will be relevant to tumor progression, but as information is accumulated for different tumor types it should be possible to identify driver mutations, and therefore to optimise treatment for patients where these changes are found. It may be that over time strategies for targeted testing for the crucial mutations will be shown to be optimal (largely reflecting current practice). Information gained from the use of WGS in tumors will at the least help in getting to that point, as well as in identifying possible new candidate drivers.

NGS, Clonal Sequencing and Finding a Needle in a Haystack

The output from Sanger sequencing represents the sum of all of the PCR-amplified DNA fragments that went into the sequencing reaction. If both copies of the gene region have the same sequence, this will be reflected in a sequence trace that shows the same base in every position. If a patient is heterozygous for a mutation then this will be apparent from a mixed pattern of bases that can usually be easily interpreted (see Figs. 7.1 and 7.2). However there are complex rearrangements which might be difficult to disentangle, and there are multiple scenarios where a secondary sequence might be present at a low level (including mosaicism for a *de novo* change, or acquired changes in a tumor that is mixed with normal tissue). NGS is clonal, which in this context means that each individual DNA fragment is sequenced in isolation. Because of the huge capacity of the new sequencing methods, each individual base of the DNA should be represented multiple times in the final output (referred to as the depth of the sequencing). This opens up the possibility of detecting sequence changes present at a very low level, and the ability to detect minority sequences against a genomic background therefore allows the use of NGS technologies to give diagnostic information where existing methods have nowhere near sufficient sensitivity. Some of these are outlined below.

Earlier chapters mentioned invasive methods for prenatal diagnosis (chorionic villus sampling and amniocentesis). The levels of sensitivity reached by NGS are such that cell-free fetal DNA present in the maternal circulation can be detected.[6] This allows for the possibility of non-invasive prenatal testing for specific mutations, detection of fetal sex, and detection of trisomies such as trisomy 21. (Because of the depth of sequencing, the additional contribution to the total cell-free DNA made by an extra copy of chromosome 21 sequences can be detected and used to confirm a trisomy.) Even if apparent positives need to be confirmed by an invasive method, the numbers of such procedures can be cut dramatically by the adoption of methods for sequencing cell-free fetal DNA.

NGS technology can also be used in the detection of circulating tumor DNA (ctDNA) in a blood sample. The value of this approach is that since tumor cells are shed into the circulation, and tumor DNA is released when these cells are lysed, information about the genetic changes present in a tumor can be gained from what is termed a liquid biopsy, without having to biopsy the tumor itself. Monitoring the level of mutations after treatment could also be used to check for signs of relapse. Potentially ctDNA testing will be used as a presymptomatic screening tool, and much work is ongoing to assess how sensitive this would be, and also to gauge the likely level of false positives—some apparent early tumor progenitors might be detected that would never have caused a clinical problem. Early results have been promising in terms of both the sensitivity and the specificity of this method.

More widely, the sensitivity of NGS methods has led to their adoption in fields such as virology, for example in looking for evolving drug-resistant strains in patients with HIV or tuberculosis, and famously in tracking COVID-19 variants.

Finding New Disease Associations

Simultaneously testing all genes in an individual rather than a panel of likely candidate genes based on the phenotype (a gene agnostic approach) can help to identify causative genes not previously associated with that set of symptoms. Such approaches rely on the sharing of genetic and clinical information

---

[6] Material from the placenta, a fetal tissue, can be shed into the maternal circulation. Fetal cells in the maternal circulation are cleared, probably by apoptosis, but this is a source of transient low levels of fetal DNA. Cell-free fetal DNA, which can be detected in maternal blood from an early stage of pregnancy, is found at varying levels at different stages of pregnancy, but generally makes up about 10% of the total cell-free DNA in the maternal circulation. Levels drop very quickly after birth—the half-life of cell-free DNA is short.

internationally, as for very rare conditions only a handful of patients will be involved, and in order to establish a new disease association, evidence must be gathered from multiple families with mutations in the same gene.

Comparative Genomics

The advent of genomic scale sequencing has opened up entirely new research fields in living and extinct populations. Because only short fragments of DNA are necessary for analysis, the standard NGS techniques lend themselves well to sequencing degraded DNA, including for example samples from Neanderthal remains some tens of thousands of years old. The reconstruction of the Neanderthal genome has shown beyond doubt that interbreeding between Neanderthals and modern humans occurred, and remnants of the genomes of more than one extinct relative are still present in our genomes today.

Of more immediate relevance, comparative genomic studies have shown that all current human populations are derived from a small group of ancestors, relatively recently, and humanity is consequently genetically very homogeneous compared to most widespread animal species. On average individuals differ at about one base in every thousand and most of these variants are of no significance. It is entirely possible to find a closer genetic match between individuals from different geographical locations than that seen between two individuals selected at random from within a population. However, there are variants with different frequencies in different groups. In most cases these are of no great significance, but are relevant in decisions on whether or not a variant is pathogenic (see Box 7.2). In a fairly recent example, genetic misdiagnoses were made in multiple American patients of African ancestry with a serious cardiac problem. Benign genetic variants were misclassified as pathogenic, based on allele frequencies from a control population that was largely of white European ancestry. While these variants were rare in the control population, apparently supporting pathogenicity, they were relatively common in other groups. This type of error can have serious consequences, in that predictive tests based on the presence or absence of these variants will be meaningless, and the true diagnosis may also be missed.

Third-Generation (Long-Read) Sequencing

In contrast to the most widely used NGS platforms, which are based on amplifying short regions of DNA, third-generation sequencing methods can read much longer stretches of sequence without amplification, routinely in the thousands of bases, and with the ever-increasing record lengths stretching into the millions (typical NGS read lengths are up to a few hundred bases). To date the long-read techniques have not been as widely used for diagnostic purposes; the main drawback is an error rate that is too high for most medical purposes (even 99% accuracy in sequencing is insufficient). However, the error rates are coming down with improvements in the technologies used, and there are specific applications where long reads are useful.

Large stretches of genomes are made up of repetitive sequences, and where the length of a repeated region exceeds the read length for NGS it can be difficult to correctly sequence these regions. Long-read platforms offer the possibility of sequencing across such regions, allowing much better coverage of genomes.

For diagnostic purposes, the ability to distinguish between genes and pseudogenes is essential for several disorders.[7] The diagnostic problem is that where pseudogenes share almost all of the same sequence as the gene of interest, PCR and other approaches will detect both, causing a problem because any mutations found could be in the gene proper, and therefore significant, or in the pseudogene, and hence irrelevant. There are ways around this involving the use either of the small sequence differences that do exist, or of flanking regions that differ. This could potentially be made more straightforward if the whole region could be sequenced in one piece, showing whether any variants were found within the gene or within the pseudogene using the context of the entire sequence of that region. This is similar to solving the problem of assigning phase to two different mutations. The phase of two mutations refers to whether they are on the same chromosome or on opposite chromosomes. This might be relevant if sequencing (Sanger or NGS) had found two different mutations in the same gene involved in a

---

[7] When a part of a genome is duplicated (a frequent and often very important event in evolutionary history) this may involve duplication of a gene or genes. In addition to the original copy, there are now one or more 'spares', surplus to requirements. In some cases, these extra copies have been gradually co-opted for a different function, and there are several examples of human gene families, often with related but crucially different functions, that originated in this way. In other cases the spare genes have degenerated – mutations can accumulate because the original functional gene copy still exists. These once-functional genes are termed pseudogenes: recognisable as derived from a working gene, and perhaps almost identical in sequence, but now a relic. There are also examples where the function has not been completely lost, and therapeutic approaches are being designed involving the up-regulation of not-quite pseudogenes in patients with a loss of the corresponding gene.

recessive condition. If the two mutations were both in the same allele, i.e. on the same chromosome, then only one of the two alleles is inactivated. If the two mutations were on opposite alleles, on opposite chromosomes, both are inactivated, confirming the diagnosis of a recessive condition. (The phase could of course be established by testing the parents of the individual in question, to show whether the two mutations are segregating together or independently.) PCR analysis and Sanger sequencing would not always give the answer, as both copies are sequenced simultaneously. Although NGS uses clonal sequencing, the fragments involved will often be too short to encompass both mutation sites. Long-read sequencing could also provide this answer, without the need for parental samples, by combining clonal sequencing with a read long enough to include both sites on the same fragment. Given the increase in the possible read length, this approach could in theory be extended to establishing the phase of any two sequence variants up to millions of bases apart.

Finally, there is clearly potential to detect and fully characterise gene fusions and other chromosomal rearrangements. As noted in Chap. 5, detection of specific fusions is essential in cancer diagnosis. A WGS approach that incorporated long reads could be viewed as a universal genetic test at the DNA level, effectively providing sequence, copy number and positional information.

Limitations to the Utility of Genome-Scale Sequencing

Given the undoubted advances described above, there is understandably great optimism about the possible uses of NGS and third-generation sequencing.[8] Sequencing costs should continue to fall, and the problems associated with the analysis and safe storage of huge amounts of data may be mitigated by improvements in informatics. (Data security is potentially a huge issue, given that a genome is uniquely identifiable.) The main bottleneck for genome-scale sequencing is at the analysis stage—generating the data is becoming increasingly routine. Better bioinformatics pipelines are already reducing the time needed for analysis, although this will continue to be a rate-limiting

---

[8] Some of the material in this and the subsequent sections was first published (in a much abbreviated form) in an article by the author entitled Genomic testing in mainstream medicine: because we can rather than because we should? *The Bulletin of the Royal College of Pathologists* (2020) no. 190 74–75. Where there is any direct overlap this material is used with permission from the Editor of the *Bulletin*.

step because of the many lines of evidence that have to be considered (see below). Above and beyond the existing ethical complexities of genetic testing, the use of routine WGS makes obtaining fully informed consent much more difficult, due to the huge potential scope for uncertain, incidental and unexpected findings, but progress has also been made in this area by current large-scale sequencing projects. Genome-scale sequencing will continue to grow in importance for the scenarios described above, and doubtless other specific applications will be developed in the context of rare genetic disorders.

However, there is an unfortunate tendency to make sweeping claims about the potential value of these technologies in mainstream medicine. The following sections set out some possible inherent limitations to the use of WGS in treating common conditions (by which is meant the major causes of morbidity and mortality, for example cardiovascular conditions, respiratory conditions, diabetes and cancer). Each of these disorders has a small subset caused by high-penetrance single gene conditions, and I will take it as established that NGS/WGS approaches have an obvious value for testing in the relevant genes. The point I will try to convey is that the case has not yet been made for the wider applicability of these approaches, and that there are good reasons for thinking that genome-scale sequencing is very unlikely to have anything like a transformative effect on the main causes of ill health. The reasons discussed below are not exhaustive, and include the genetic complexity of common disease, the variability of the human genome, other biological (including epigenetic) confounding factors, and probabilistic arguments, although inevitably these categories will overlap.

Confounding Factors: Complexity of Common Disease

Common medical conditions, for the most part, are not caused by highly penetrant mutations in single genes. Susceptibility to common conditions overwhelmingly involves multiple loci, each individually contributing to a small proportion of the increase in susceptibility, and interacting with the environment, the effects of variants in other genes, and epigenetic factors. The potential complexity of these multiple interactions is enormous, and complete knowledge of all of the contributing factors is unobtainable. For common medical conditions with a genetic component such as diabetes, there are rare monogenic types. However, these only account for a small fraction of the total. In the majority of cases, genetic susceptibility to such conditions has been shown to be spread across hundreds of genomic regions, and in no case has all of the heritability for common conditions been accounted for—there are plenty more variants to be found, none of which will have a large individual impact. To take the example of diabetes, in addition to the complex and diffuse genetic basis of the condition in most patients, development of clinical symptoms is inextricably linked to the well-established modifiable risk factors of obesity, sedentary lifestyle and hypertension. Testing all of the known genetic associations would, in most cases, make no difference to the diagnosis or treatment of the patient (and note that these are associations rather than causes—the methods used to define these regions give no information about any mechanism by which the susceptibility is increased by some fraction of a percentage point). Lifestyle advice would not differ either, and it is difficult to see what WGS would add to any benefits that could be provided by targeted polygenic scores.[9] Even if a whole genome analysis predicted (for example) a 15% additional risk of diabetes on top of the population average would that actually change behaviour? If a lower than average risk was found, would it make the individual less careful? Either way, as things stand the potential impact of WGS on common disease looks more marginal than transformative.

Confounding Factors: Lots of Variation, Many Rare Variants

Variability of the human genome is a major confounding factor. Any genome-scale test will detect a daunting total of rare and potentially significant inherited and *de novo* variants and, as mentioned earlier, allele frequencies can differ significantly between populations. Because most variants are individually rare, it is difficult to collect evidence to allow final confirmation or exclusion of pathogenicity, and perhaps the current biggest single limitation to the utility of genome-scale sequencing is the difficulty of deciding whether or not a sequence change is pathogenic, or is a neutral variant. The same limitation applies to variants detected by Sanger sequencing or NGS of a defined

---

[9] Polygenic scores are derived by combining the effects of multiple common genetic variants and using the combined score to give a modified risk of a condition such as raised cholesterol—for most individuals tested either slightly higher or slightly lower than the population average. Such variants are identified by genome-wide association studies, where multiple patients are tested to see whether those with a particular condition are more likely to share particular variants than would happen by chance. This establishes at best a statistical association of a variant with a disorder; correlation is not causation. Polygenic scores will be further refined as a way of clarifying risk, but the impact is likely to be incremental rather than transformative. To be told you have a higher than average risk of high cholesterol may persuade some to modify their behaviour, but we all already know that we shouldn't drink beer and eat pies to excess.

panel of genes, but because these assays are more targeted, there will be fewer variants to classify, and the variants will be in genes that have a higher prior probability of being causative.

There are guidelines in place to help with classifying these variants, the aim being that all labs would come up with the same classification. It is clearly a problem if two different centres come to different conclusions regarding the same change. Guidelines are constantly evolving, and generally becoming more complex. As many lines of evidence as possible should be taken into account and weighted appropriately. Some of the basic principles are laid out in Box 7.2.

The aim is to classify any sequence change into one of five classes, ranging from definitely disease causing to definitely neutral. The five classes are pathogenic, likely pathogenic, variant of uncertain significance (VUS), likely benign and benign. These distinctions matter to the patient: the first two are considered as actionable, so for example a patient with a pathogenic or likely pathogenic change in a cancer predisposition gene could be offered additional screening, testing in other at-risk family members and perhaps prophylactic surgery. Benign or likely benign changes would not usually be reported. The guidelines tend to be conservative, and unfortunately many variants end up classified as a VUS—of little help to the patient and their family. This reflects the gaps in our current state of knowledge, including the inevitable lack of information about a change that might never have been seen before. All changes are subject to review in the light of new knowledge, and it is a significant headache for laboratories to keep on top of reviewing all of the variants previously reported. However, review is essential—the existing guidelines are known to be an imperfect tool. There are well-established pathogenic mutations and benign variants that would have been misclassified by an application of the rules as they stand. It is of course the case that as knowledge increases, and analysis tools improve, we will develop a better understanding of which variants are pathogenic and which are benign (and it is also the case that once a genomic sequence has been obtained for a patient, these data can be revisited in the light of new information). Even so, the complexities of human genetics will always cause problems for analysis. For example, the majority of genes produce multiple different mRNAs (transcript isoforms) through different patterns of exon splicing during transcription, and through the use of alternative start points for the transcription process. The result is that the human proteome, the sum total of all proteins found in human tissues, is significantly larger than the gene number—most genes that do code for proteins produce multiple different products. Whether or not a sequence change is pathogenic may depend on whether that change affects a biologically relevant transcript isoform, which in turn is likely to be tissue specific and to be modulated by environmental factors.

Box 7.2 Assigning Pathogenicity to Sequence Changes: Principles

When trying to determine pathogenicity for a sequence change, one obvious factor to consider is how rare that change is. If the population figure for a variant is higher than the relevant disease frequency in that population, then (with allowances for partial penetrance) that variant cannot be causing the disease. There is a huge and growing body of well-curated data that allows population frequencies to be easily checked. In practice common variants by definition are well known, and can be automatically filtered out as part of the analysis process. Among the important constraints to the value of this step are some that are temporary (limited data, especially for certain populations and ethnic groups, but this will improve) and some that are more fundamental. Most variants are rare, and ultimately it will always be difficult to gather sufficient information based on population frequency alone to classify such rare variants. More weight is given to evidence that a variant has been seen in multiple affected individuals, but is rare or absent in the general population. However, caution should be applied when relying on previous reports of changes found in association with a genetic disease—correlation is not causation.

The consequence of the sequence change is one of the strongest indicators of pathogenicity. If a stop codon is introduced (either directly as a nonsense mutation or as a result of a frameshift) this is highly likely to be damaging. The same applies for changes at the start and end of the introns, which can abolish correct splicing of the mRNA. Missense mutations may have a more variable effect, depending on how critical that particular amino acid is to protein function. Synonymous mutations (as most amino acids are specified by more than one codon, some DNA changes won't alter the amino acid sequence) are usually benign, but in some cases might affect a splicing or other regulatory signal. There are programs that can be used in predicting how a missense change might affect a protein or how a synonymous change might affect splicing, but at present (and although improving) these are nowhere near good enough to give conclusive evidence in most cases.

Although finding a relevant change in the DNA of a patient is the goal of genetic testing, the real importance is in how this affects the function of the gene product. Ideally, the pathogenicity of a change should be assessed by looking at functional assays. In practice these are very rarely available. Of the countless millions of potential genetic changes, only a small handful have been investigated because of the difficulties in designing specific assays (and even then these are usually performed in vitro (in artificial conditions) and therefore the relevance to a living system can always be doubted). Once again, knowledge will improve as it becomes easier to artificially construct mutations to test, but very rare conditions are unlikely to be included in such analyses.

If a change occurs *de novo* in an affected individual with unaffected parents this can be used as evidence that supports pathogenicity, but once again is not conclusive—we all carry *de novo* mutations, mostly without consequence. If the change is present in a gene known to be associated with the relevant symptoms this would carry more weight (especially if tests had failed to show any other mutations in that gene or other genes associated with the condition).

Depending on the family structure, it might be possible to assess pathogenicity by looking to see whether the sequence change is seen in all affected family members. It could of course happen by chance that a benign variant segregates with the disease, so the value of this evidence depends on how many affected family members there are. If the change is present in an unaffected individual or absent in an affected individual this would count against pathogenicity (although as usual there are complexities, respectively around penetrance and the possibility of a family member coincidentally having similar symptoms).

Confounding Factors: Complex Metabolic Networks

Results from genome-scale studies suggest that healthy individuals carry a surprising number of variants that on the basis of existing tools would be predicted to be pathogenic, but apparently (for that individual in their current environment) are not. Metabolic networks are known to be resilient: there is often a high degree of flexibility in the system, alternative routes to get to a given metabolite. A high degree of genetic redundancy has been found even in organisms with a relatively small number of genes.[10] Publications on gene function routinely include diagrams showing interactions with other genes in networks of staggering complexity. The possible total number of permutations for gene interaction is enormous, even before environmental effects are factored in. It is entirely possible that the same variant in a gene could have different consequences in different individuals depending on the genetic background—benign or even advantageous in some genomes, damaging in other circumstances.

The question is not just whether a variant is pathogenic, but whether it is pathogenic in a particular context. High penetrance single-gene conditions represent one end of a spectrum, and although these may be the conditions that come to mind to those currently involved in genetic testing, common conditions predominantly lie at the other end of the spectrum.

---

[10] A painstaking project (introducing deletions knocking out all of the 6000-plus protein-coding genes in yeast cells one at a time) has shown that many of the yeast strains completely lacking a gene are still viable.

Epigenetic Regulation: A Further Level of Complexity

The concept of epigenetic control (factors above and beyond the DNA sequence influencing gene expression) was introduced in Chap. 4 in the context of X-inactivation. DNA methylation as a means of silencing genes is a critical factor in normal development and response to the environment, and changes from the usual pattern of methylation are implicated in inherited disease and in cancers. There is also a small but increasing body of evidence supporting the possibility of acquired epigenetic changes being passed on through several generations—it remains very much a live research area to flesh this out and to establish what the mechanism might be. In addition to acquired changes, there are also regions of the genome where, irrespective of the DNA sequence, patterns of gene expression differ depending on whether that region is inherited maternally or paternally. This difference in expression is marked by DNA methylation. Although parental origin has no effect in the majority of cases, some genes are only expressed on a maternal allele, with the paternal allele silenced by methylation, and for other genes the opposite applies. This is known as imprinting (see Box 7.3).

Epigenetic changes are very much a feature of normal development. Whether or not acquired epigenetic changes can be inherited, such changes have been shown to have long-lasting effects on gene expression within an individual in response to environmental factors. This is another level of uniqueness for all humans—in addition to the unique (other than in identical twins) genome, the pattern of gene expression reflects unique life histories. The functional genome is not carved in stone, even if the primary sequence is largely fixed, but is in constant dialogue with a changing environment. The extent of the role of epigenetics is only just beginning to be appreciated, and differences in gene expression due to this level of control are not captured by standard sequencing; an inactivating mutation (or a gain of function mutation) in a gene silenced by imprinting or an acquired epigenetic change will have no additional functional consequence. Although such factors can perhaps be safely ignored in assessing pathogenicity for single gene high penetrance conditions, epigenetic effects are likely to be hugely influential in the context of common multifactorial disorders such as diabetes. Any expectations for the utility of WGS in the management of common conditions will have to take into account the impact of epigenetic factors.

Box 7.3 Imprinting and Genetic Disease

The evolutionary origin of imprinting is still something of an open question, but the implication is that for imprinted regions of the genome, inheriting both a maternal and paternal copy is essential for normal functioning. Where a pregnancy occurs due to an entirely maternal or entirely paternal contribution (due to doubling of the genetic material of an egg or sperm) normal development is impossible. Although for most genes inheriting two paternal or two maternal copies is not in itself problematic, a global lack of biparental inheritance for imprinted regions is not compatible with life.

The imprinting pattern is maintained throughout standard cell division and is therefore stable in somatic tissue. In the germline the pattern is erased, then re-established as either a female pattern in all eggs, or a male pattern in all sperm. Occasionally, most often due to the consequences of an error in the way that chromosomes have segregated during meiosis, an individual will inherit two copies of one chromosome from the same parent (the technical term is uniparental disomy, UPD). In the absence of imprinting (and unless the UPD by chance unmasks a mutation in a gene for a recessive disorder on that particular chromosome) this may not matter. However, there are some chromosomes where maternal or paternal UPD does cause clinical problems. These chromosomes have imprinted regions where maternal and paternal genes are differentially expressed, and the clinical features therefore may differ depending on whether paternal or maternal UPD has occurred. Since the difference is associated with differential methylation, which can be detected by a variety of assays, the parental origin of the chromosomes can be established and a diagnosis made.

UPD is one of the underlying causes of two genetic disorders, Prader-Willi syndrome (PWS) and Angelman syndrome (AS). These two conditions are clinically quite distinct, but the underlying error can be exactly the same in both (including most commonly the deletion of a large region of chromosome 15, as well as UPD of chromosome 15). The difference lies in the parent of origin. With a deletion of the critical region on a paternal chromosome or maternal UPD (in both cases losing the paternal contribution) PWS is the result. A maternal deletion or paternal UPD gives rise to AS. Different genes are effectively lost because they are either silenced on the maternal copy (many of the genes in this region) or on the paternal copy (one specific gene). None of the crucial genes is expressed on both, so lack of either a maternal or a paternal copy of the region is pathogenic, and because effectively different genes are lost, different syndromes result.

The Risk of False Positives

Where the prior probability of a positive result is low, even an apparently highly specific test will generate an unacceptable level of false positives. This is straightforward when applied to a simple yes or no test (see Box 7.4) and arguably also applies to WGS when testing a large number of genes for a possible single gene disorder (in this case defining a false positive as a variant that cannot be excluded from being pathogenic but which is in fact not damaging, any such variant in any gene being a false positive in this sense). The situation is even more difficult when testing for common conditions, since a whole genome screen would combine a low prior likelihood that any individual gene will be contributing significantly to the clinical phenotype, the near certainty that this contribution will depend on the environmental, genetic and epigenetic context, and the inevitability of detecting a large number of variants of uncertain clinical significance. If you throw a dart at a single bull's eye you can assess the accuracy of the throw, as opposed to throwing a handful of darts at a wall and then drawing targets around the darts where they stick.[11] In the context of testing for rare inherited conditions, the WGS approach is at least potentially aiming at a single bull's eye. In the context of common multifactorial conditions, the hundreds of targets are ill defined at best.

Box 7.4 False Positives in Diagnostic Tests

The false positive burden for a test does not just depend on the intrinsic qualities of the test but on how common the disorder in question is. Suppose there was a test with 100% sensitivity for detecting COVID-19 and 99% specificity (i.e. all true positive cases would be detected, with a 1% false positive rate, which is better than a lot of the real-life tests in use). If in the population being tested, 1 in 100 individuals are currently affected, and 1000 individuals are tested, 10 true positives and 10 false positives would result—only a 50:50 chance that you are actually positive if the test says so. If the population frequency is 1 in 1000, the same number of false positives would be found, but only one true positive—false positives outnumber true by 10:1. These numbers don't look good if applicable to genome-scale tests in the context of single gene testing, let alone testing for common disorders. A Bayesian approach tells us to be careful where the prior probability of an event is low.

Will WGS Improve Outcomes for Common Disorders?

For patients with a rare genetic disorder, getting a diagnosis has importance in itself, as well as in potentially clarifying recurrence risks, even if no specific intervention is possible. For common disorders, diagnosis is not

---

[11] The analogy of the moving bull's eye was used by the philosopher Simon Blackburn in his book investigating the notion of truth.

usually the issue. For WGS to have a value in mainstream medicine, it would have to be as part of a route to improved treatment, or as a predictive tool. However, even in the unlikely event that all of the variants contributing to a common condition could be unambiguously identified, there is limited evidence that this would allow any interventions (or predictions) beyond what is already possible. In the case of cystic fibrosis, a relatively common and well-understood single-gene disorder, specific treatments based on genetic information have only recently become available and these are very expensive. Similar gene-specific interventions are very unlikely to be developed for multigenic conditions.

When requesting a laboratory test clinicians are asking a clinical question on behalf of a patient. In the context of genetics, the question might be whether or not the patient is affected with, or a carrier of, a specific genetic disorder. In an ideal world, the clinician and the patient would want a clear and unambiguous answer, one that will suggest treatment options, and they want it as quickly as possible. More information than is needed to answer the question is not necessarily helpful. In some of the circumstances detailed in the earlier part of this chapter, WGS represents the best way of answering that question. For mainstream medicine, where a genetic test is important (as it is for many cancers), a more targeted test will often be faster, cheaper and just as useful. Using a specific test also reduces the risk of unexpected, unclear or incidental findings (including discoveries about familial relationships, carrier status for recessive disorders, or mutations in susceptibility genes for late-onset disorders).

Much has also been made about the possibility of genome testing facilitating personalised medicine in terms of the individual response to drug therapies (the field of pharmacogenomics). There are already some examples of genetic testing being used routinely in gauging how well a patient will metabolise a particular chemical. To a high degree differences in drug response are heritable, so it makes perfect sense to use this approach when the relevant variants are known, to avoid side effects and maximise the therapeutic value. However, an investigation that requires analysing a small number of variants in a short time frame again lends itself much better to a simple targeted genetic test.

Genomic Testing in Mainstream Medicine: Because We Can Rather Than Because We Should?

In contrast to the clear improvements for rare genetic diseases, there is yet to be a case made for the value of WGS in the management of common conditions. The complexities listed above suggest that the inherent limitations on the utility of genomic sequencing for multifactorial conditions are real and significant. Facile promises of imminent personalised medicine across the population ignore the complexity of the interactions within the genome and between the genes and the environment. In general, advances in genetics often give rise to unrealistic expectations. Announcing plans to sequence impressive numbers of genomes (usually some multiple of 100,000) has become a common event, coupled with claims for a transformative effect on healthcare.[12] A transformation of healthcare implies tackling common disorders, and other than for cancer, there are as yet no good examples of the use of WGS in informing treatment for the big causes of mortality and morbidity. Even in the case of cancer, it is to say the least an open question as to what degree there are benefits in testing at genome scale, rather than using rapid and cheap targeted testing for the handful of driver mutations that are important for diagnosis and treatment. You don't have to be of an overly sceptical frame of mind to wonder whether these projects owe more to political imperatives than to evidence-based medicine. There needs to be an informed debate before the introduction of large-scale genome sequencing in mainstream medicine, especially if (within finite resources) this is at the expense of more targeted genetic testing.

---

[12] For example, the UK NHS Genomic Medicine Service website describes a plan to "offer whole genome sequencing as part of routine care", and to sequence 500,000 genomes by 2023/24 to help "transform healthcare for maximum patient benefit" (https://www.england.nhs.uk/genomics/nhs-gen omic-med-service/, accessed 16 April 2021). This all sounds wonderful, but is going way beyond the evidence in the absence of large-scale studies looking at the benefit of WGS in routine care, as opposed to testing in patients with rare genetic diseases. The question of resources also tends to be better addressed for the initial sequencing capacity rather than in dealing with the huge data sets acquired—analysing 500,000 genomes is not trivial in terms of scientific input.

8

DNA Testing: Pulling the Strands Together

Summary In the final chapter, some of the themes that have been discussed or hinted at earlier in the text are revisited, and some general observations are made on laboratory genetics as a scientific discipline. An overview is attempted of some of the main professional and ethical responsibilities for diagnostic geneticists, touching on issues of providing benefit to the patient, consent, autonomy and avoiding harm (potential harms including the risk of genetic labels, the downside of open access to genetic testing, and the possibility of misinterpretation of results). Inevitably, some of the questions raised remain open; in complex ethical areas, a definitive answer isn't always available. A broad definition of the utility of genetic testing is needed (encompassing more than solely facilitating management of a condition). The final parts of the chapter are concerned with the uses and misuses of genetic language and the need for a better societal understanding of the benefits and limitations of genetic testing.

Diagnostic Genetics and Ethical Principles

Looked at in one way, being a human is entirely genetic—we develop as we do because our genes determine that we are human. Looked at in another way almost nothing is completely genetic—anything of interest about a person is sculpted by the culture they are brought up in and their physical environment. Being involved in genetic testing can lead to an over-emphasis on the role of genes, perhaps understandably, as the daily work involves looking for

© The Author(s), under exclusive license to Springer Nature
Switzerland AG 2022
D. Bourn, *Diagnostic Genetic Testing*,
https://doi.org/10.1007/978-3-030-85510-9_8

mutations that affect single genes, and thereby cause genetic disorders in a more or less deterministic way. Effectively, the process of testing for specific genetic changes has to follow the classic reductive scientific path of looking at one element in isolation (see Box 8.1). This carries certain risks. In theory, it would be possible to spend an entire career in a diagnostic genetics laboratory without thinking too much about the consequences of the tests you are carrying out. However, although laboratory staff in the main do not have much direct contact with patients, close interactions with clinical colleagues should help in ensuring that everything is done with the benefit of the patient in mind. The principal motivation of those working in laboratory medicine is to apply their knowledge for the general good.

Diagnostic genetics is no different from any other area of medicine in that practitioners should always adhere to the basic principles of medical ethics. Ethical awareness around consent, patient autonomy, avoiding harm and benefitting the patient is (or should be) a part of any aspect of medical treatment. Equity is also important. Access to genetic testing has always varied both within and between nations, as a consequence of differential resources, different healthcare systems, and other historical and social factors. Where a technology is beneficial it should be made available as widely as possible. In practice new and initially expensive technologies, particularly those addressing health problems that may be chronic rather than acute will always have a greater initial take up in wealthy regions. However, once new techniques are established and proven useful, there is at least the potential that the knowledge can be shared rapidly, and disseminating such information as widely as possible should be a priority. The community also has a responsibility to insist that new technologies are used because we should, rather than because we can; any changes in practice should be evidence-based, rather than driven politically in eye-catching areas like genomic testing.

As well as the onus on those working in the field of genetics to follow general ethical principles, there is a responsibility to engage with some of the specific ethical dilemmas raised by the practice of genetic testing. Problematic areas include the power of such testing to predict outcomes, the intrinsic nature of genetic information (in that any finding may have important and direct consequences for relatives), and more recently the introduction of testing that can give rise to unexpected, unwanted or uncertain findings. The succeeding sections include some tentative suggestions as to how to engage with these more involved questions.

Box 8.1 What is the Scientific Method?

It is surprisingly difficult to define science. No simple description exists that fully captures the field. A common sense definition of science as an experimental discipline, working under carefully controlled conditions and altering one feature of a system at a time, recognises an important part of the process. However this ignores the role of underlying assumptions based on the existing body of knowledge which inform how a result is interpreted, and of course it is true that there are limitations to the degree to which our imperfect minds, senses and tools can actually grasp what is true about the external world. Nevertheless, for day-to-day work, abstract speculations, however interesting, have to be set aside in laboratory medicine, where the requirement is to come up with results to inform patient care in a reasonable time frame. Most practising scientists (assuming that they consider these areas at all) are probably comfortable with the idea that science is provisional, to be reviewed in the light of further information. A scientific theory has to be falsifiable in the sense laid out by Karl Popper. If there is no possibility of finding evidence that could count against a theory, no way in which it could be disproved, this should not count as science. Successful scientific theories are those which make useful predictions and have thus far survived any attempts at disproof. This implies that any piece of science could become redundant, although in practice there are well-established theories that will probably remain useful indefinitely, at least as a subset of the broader and deeper theories that have subsumed them. As should be expected, diagnostic laboratory genetics is based on concepts that have been thoroughly tested and have so far survived these tests. The pragmatic view would be that what we have now works, and we'll keep using it until something better comes along. In practice results can be measured against the real world in important ways: if a prediction is made that an individual will not develop HD or an inherited cancer, and they go on to do so, this will inevitably come to light. Ideally (and to a high degree in practice) errors due to human or systematic fallibility are kept to a minimum, and errors due to a fault in the basic science are negligible. Crucially, any competent laboratory should come up with the same answer for a given test on the same patient (and laboratories are subject to external comparisons to ensure that this is the case). Where diagnostic genetics does have to contend with uncertainty this needs to be made clear, using the appropriate probabilistic statements and figures. The Bayesian approach is well suited to giving the best answer possible under limited information, taking a known or estimated initial risk and modifying that risk in the light of all of the relevant genetic (and other) information that can be gathered.

Consent in the Genomic Era

Informed consent should be obtained for any clinical procedure, including laboratory testing. This is relatively straightforward where the consent is to a test giving a simple yes or no answer to a clinical question, most often asking whether or not a patient has a specific disease. The patient is already ill, and wishes to know the cause, ideally to direct treatment but at least to give them information. Consent for predictive testing is more complex, but at least this still involves a specific test, with a binary outcome. If the test in question is a whole genome sequence, with many possible findings and many unknowns, how can meaningful informed consent be given? This problem has had to be addressed for the various genome projects carried out worldwide. Do patients want to know about adverse findings that are unrelated to their current clinical symptoms? Do they want to know their carrier status for recessive conditions? Do they want to know about predispositions to late-onset cancer syndromes, and have to make decisions on screening or prophylactic surgery based on a probability rather than a certainty? How about findings of variants with an unknown effect? Given these uncertainties, it is difficult to see how consent can be fully informed for such a new technology, and further work needs to be done with patient groups to ensure that consent forms successfully balance the need to convey complex information with the need to be intelligible. It is already difficult for genetic counsellors to conduct discussions around how information should be shared within a family for single gene conditions, and this becomes even less tractable with the increase in information associated with genome-scale testing. Encouraging overly raised expectations should be recognised as a particular risk. Some of the language used regarding the transformative potential of whole genome sequencing in mainstream medicine goes well beyond what is justified by the findings to date, and promising families answers and treatments that may not be forthcoming stores up problems for the future.

Making Genetic Choices

Respecting the decisions made by a patient, in the light of the information given to them, is a central pillar of medical ethics. Where information is uncertain or probabilistic, how is patient autonomy best served? How much information should be given around the grey areas in such a fast changing field? In the case of genetic testing, there may be tensions in considering the autonomy of different individuals within the family where there is a conflict. For example, should the autonomy of the mother be placed above the autonomy of an unborn child? No human being should be treated as just a means to an end, but there may be situations where the rights of two individuals to be treated as an end in themselves are incompatible. If prenatal diagnosis and termination of pregnancy is made illegal, is the mother then being treated as a means to an end in having to give birth to a child known to be at risk of having a severe genetic condition? Where terminations are legal, is the affected fetus being treated as merely an obstacle to the parents having a child without such a condition? It is considered that testing with the sole purpose of discovering carrier status should not be carried out in minors out of respect for their autonomy; the view is that they should make an informed choice when reaching adulthood as to whether they want to know their carrier status. As this knowledge is mainly of importance for reproductive decisions, the rationale is that gaining this information is of little benefit to the child, and might put a harmful label on them as a carrier. However, this runs a risk that perhaps many years down the line the lack of this information affects the ability of that individual to make an informed decision (there is no guarantee that the family of that individual will make them sufficiently aware of any risk, although the same might apply even if the test is carried out in childhood). Where does the burden of respecting autonomy lie, in the imperative not to give unwanted information to the child or in the imperative not to withhold information that might be useful to that same individual as an adult? These are not areas where there is any simple answer, and unique cases and difficult conflicts will always occur.

Compartmentalisation on Genetic Grounds

Genetics has perhaps the most problematic history of all scientific disciplines. Although most of the abuses involved in eugenics and social Darwinism predate the discovery of the role of DNA and the era of modern genetics, the story of the twentieth century includes innumerable examples of extreme human suffering caused by a misuse of what were thought at the time to be scientific principles. The genetics profession has been co-opted on the wrong side of history more than once. Individuals and whole groups have been defined as subhuman on a variety of spurious grounds, and once someone is characterised in this way it gives licence to treat them in ways that would not be allowed for 'normal' citizens. It should go without saying that even where someone has some damaging change to their DNA (which in varying degrees probably applies to all of us) this should not alter their right to be treated with the respect due to any human being. The use of any genetic deviation from the norm to justify treating an individual differently is inherently dangerous (and in practice the definition of the norm is invariably to some extent arbitrary).

Suppose that it were found that a particular gene or combination of genes predisposes to criminal behaviour. How should this knowledge be used? Should everyone be tested, and those with such an inheritance subject to surveillance or preventative incarceration? To what degree will your genetic inheritance alone determine your actions? There is some research which indicates that (in a very small number of families) a rare single gene defect could predispose to criminality, but in general any complex human behavioural trait will be influenced by an intricate mix of genetic and environmental factors. To use a very familiar example, it has often been pointed out that there is in fact conclusive evidence that a known combination of genes shows an association with criminality, and particularly with violent crimes. The vast majority of those convicted of offences involving violence are known to have this combination of genes, usually referred to as the Y chromosome. There is clearly a huge problem of male violence, which needs to be acknowledged and the appropriate preventative and rehabilitative measures put in place, including early education. Coming at this from a genetic angle adds nothing, and risks the damaging perception that this behaviour is determined or even 'natural'.[1] There have been historical suggestions that males with two Y chromosomes (XYY males, a condition that occurs in around 1 in 1000 male births, and usually goes undetected) are overrepresented in prison populations (the original reports date back many years and have always been controversial). Using genetic information of this type to inform public policy would be deeply unhelpful—there are so many other factors, and the effect of environment on criminal behaviour is apparent (and given the political will, much more amenable to change than are genomes). Slippery slope or thin end of the wedge type arguments can be overused.[2] However, there are real risks inherent in the extension of genetic compartmentalisation, and genetics professionals have a strong responsibility to subject any such moves to rigorous scrutiny, and to call out any misuse of the science. This applies equally to drawing attention to the unjustified uses of genetics described in previous chapters, in making arbitrary categorisations of race and in general attempting to draw clear dividing lines on what is in reality a continuum.

---

[1] Even if this were true, the existence of a set of behaviours in nature tells us precisely nothing about the desirability or morality of those behaviours in human society. Many things that are the case would not be in an ideal world—as David Hume realised, you can't get an ought from an is.

[2] An increase in the road speed limit does not lead automatically to total derestriction, and a decrease does not lead to mandated walking pace. Any limit is arbitrary to some extent, and it should be perfectly possible to agree to a common-sense consensus in most areas. In the present context, the risks of increasing compartmentalisation on the grounds of genetic differences can be mitigated by a better understanding of the dangers and contradictions of such an approach.

Commercial Access to Genetic Testing

In most healthcare settings, laboratories that carry out genetic testing are well regulated, and the staff are fully aware of their duty to the patient. Such laboratories will work closely with clinical colleagues to ensure that testing is carried out in an ethical way. Direct-to-consumer (DTC) testing may not always be constrained to quite the same extent, although clearly having to abide by the relevant legislation where such testing is permitted. However, legality and ethics do not always map perfectly on to each other, and there are available DTC tests that will give you your carrier status for severe recessive conditions, and will look for mutations in genes that can confer a high risk of (for example) breast cancer. A number of other examples are given in the sections on individual genetic disorders. Receiving such life-changing information without the context of full genetic counselling can have serious consequences for the individuals concerned. The downstream confirmation of the test results and the genetic counselling will then (in nations with state health provision) fall to the national healthcare provider at the taxpayer's expense, while the profit goes to the company providing the test. It would be difficult to prevent the use of such widely available technology, and many would support the idea that individuals who choose to spend their money on a genetic test should be allowed to do so, despite the risk of harm. It is less clear why this should be allowed to place a burden on state health systems, which are then effectively subsidising the process.

Acknowledging Uncertainties and Avoiding Error

A genetic test can't always offer a clear and decisive answer, even when testing single genes. Uncertainty may be due to intrinsic limitations in the sensitivity and specificity of the test (although specificity is generally less of a problem when looking at discrete sequence differences). Where it is not possible to look at all mutations there is always the risk that a rare, undetected mutation might be causative in that patient. In contrast, if whole genome sequencing is used (which by its nature should pick up any sequence changes) the uncertainty may be around which (if any) of the huge number of variants detected is responsible for the clinical problem.

The unspoken assumption behind much genetic testing is that mutations in a single gene are likely to be responsible for the symptoms. In reality most disease, where it has a genetic component at all, will have a far more complex aetiology involving multiple genes, interacting with environmental factors. At present we do not have sufficient knowledge to untangle these networks, and although knowledge is always improving it is essential that the benefits of genetic testing are not overstated. Failure to live up to the hype surrounding whole genome sequencing will lead to a loss of confidence in promises made to the public. Large claims for the benefit of scientific advances require appropriate factual justification rather than political wishful thinking.

Any human enterprise is subject to errors and biases. For the most part laboratory medicine has sufficient checks in place to reassure the public that errors are rare, and will be systematically investigated when they do come to light. Often in genetic testing internal inconsistencies in the results will prompt careful investigation for a possible mistake.[3] However, other errors could easily be missed (swapping a sample from a patient at risk of a genetic condition with that from another patient is quite likely to lead to a false negative result, given that a sample from a patient referred for a different reason will almost certainly not carry the relevant mutation). The possibility of an incorrect result is minimised by both internal measures (adherence to policies and procedures) and external measures (accreditation to high standards by mandated inspections).

The nature of genetic testing for rare inherited disease and acquired changes in the genome makes it inevitable that unusual (and often unique) findings result. There may be no precedents that allow a secure interpretation of such results. Conversely, there is always the risk of relying too much on a particular line of interpretation that has worked in the past, and it is therefore important to be aware of confirmation bias—giving more weight to evidence that supports your pet theory than to evidence that would falsify it.[4] This is where the value of science as a collaborative and iterative process is helpful; as well as discussion with local colleagues, conversations with the wider community allow the pooling of knowledge to give the best chance of a correct interpretation. Science should keep its practitioners honest—as

---

[3] For example the results of testing may show an apparently male sample for what should be a female patient or vice versa, there may be apparent non-inheritance of familial variants where this should be impossible, there may be a discrepancy between clinical features and mutation presence/absence, and so forth.

[4] In an autobiographical fragment Charles Darwin wrote that he had "during many years followed a golden rule, namely, that whenever a published fact, a new observation or thought came across me, which was opposed to my general results, to make a memorandum of it without fail and at once; for I had found by experience that such facts and thoughts were far more apt to escape from the memory than favourable ones."

Thomas Huxley said, a beautiful theory can always be slain by an ugly fact—and especially where patient results are concerned it is unacceptable to allow an interpretation of a result to go beyond the evidence.

The Value of Genetic Testing

The clinical utility of a test refers to the value of that test in the overall treatment of a patient. It would be unethical to carry out any medical test that did not provide a benefit for the patient, and genetic testing must satisfy this criterion in the same way as any other discipline.

A standard definition of the clinical utility of a medical test would emphasise the degree to which the result might inform specific management or treatment options. If the output of genetic testing is judged against this narrow definition, much of the practice might fail to qualify. A broader definition of utility is needed to capture the value of much genetic testing, which consists in giving patients and their families information, for example by allowing a diagnosis which would otherwise be unavailable (particularly for some of the very rare conditions). Just knowing what is causing your symptoms or those of your child is important, even in the absence of any intervention. Knowing the implications of that diagnosis also allows for a measure of control, whether that be in planning reproductive options or having access to more screening.

While the potential risks of and uncertainties of inappropriate genetic testing should be acknowledged, when used properly genetics is a valuable diagnostic and predictive tool. As described in the previous chapters there are many instances where genetics can give a definitive answer. The tests conducted routinely in diagnostic labs are precisely those where (for the most part) the answers will be clear in that a diagnosis will be confirmed or excluded (or at the very least the probabilities of the diagnosis shifted significantly one way or the other). Even where a definitive diagnosis can be made on clinical grounds or by using other laboratory tests, a genetic result may be of great value in giving information about the risk to other family members. Genetics is uniquely useful in the scope for carrier and predictive testing. However, in some cases the genetic result will be only one piece of evidence among many—useful, but not the only tool in arriving at a diagnosis or prognosis. An obvious example is for cancer, where the genetic component of testing constitutes just one part of a battery of tests. (Although as described in Chap. 5 this has changed to some extent with improvements in testing, and even more so with the development of specific cancer drug therapies, based on the underlying driver genetic events.)

It is still the case that a genetic diagnosis does not usually allow a specific and targeted treatment. However, this situation is gradually improving. As discussed in Chap. 3, there are now very successful treatments for the underlying gene defects for most cystic fibrosis patients, and a specific treatment has recently been licensed in some nations for spinal muscular atrophy, another severe autosomal recessive disorder. It is hoped that such personalised medicine will be extended to designing specific drug regimens for individuals based on typing genetic variants involved in drug metabolism. The field of pharmacogenomics offers the promise of tailoring drug treatment to the patient depending on their genotype, rather than a one size fits all or trial and error approach to dosage.

The Language of Genetics: Uses and Misuses

The phrase 'genetic language' is traditionally used in descriptions of the genetic code as an analogue of a written language. Increasingly, however, genetic terminology and gene-related metaphors are being employed more or less accurately in popular culture—a broader meaning of 'genetic language'.[5] References to DNA feature prominently in media coverage of medicine and forensics, and genetic testing was dissected minutely in news and documentary coverage of the COVID-19 pandemic. Some of the language is almost always used incorrectly. When reporting on a new finding about the association of a gene with (for example) bipolar disorder, this will often be referred to by journalists as a gene for bipolar disorder, rather than the more nuanced, less snappy but correct description as a gene where certain variants may (in combination with other genes and many environmental factors) increase the predisposition to bipolar disorder above the average for the population. In a similar vein, references are often made to an individual 'having the gene' (for example in a family where a *BRCA1* mutation is present). Everyone 'has the gene' but some individuals have a version of that gene carrying a

---

[5] As a case in point, the phrase 'the selfish gene' has been immensely influential, in ways that show both the power and the pitfalls of metaphor. The basic theme of Richard Dawkins' most famous book (outlining a gene-centred view of evolution) is succinctly captured in the three words of the title. However, some of the less careful critics of the work interpreted this to mean that genes are in a literal sense selfish, apparently a wilful misreading of the book. Dawkins' ability to express a concept in a way that moves genetic language into popular culture is exemplified by his coining (also in *The Selfish Gene*) of the term 'meme' as a 'unit of cultural transmission', analogous to the gene as a unit of genetic transmission. Appropriately, the term has found a new environmental niche in describing successfully propagated concepts in social media.

damaging sequence change. As well as these unhelpful but relatively benign misuses of gene-related terms, there are more problematic and damaging ways to abuse the language of genetics. Eugenic ideas of a pure race and the evils of racial mixing have no basis in fact, (and one indisputable finding from genome research is that the term race as commonly used has no biological basis whatsoever).[6] Unfortunately, the abuse of scientific sounding terms and poorly understood ideas on evolution and survival of the fittest have a long history, and not just in the nations with a particularly bleak reputation for ethnic discrimination—there are European states with current impeccable liberal reputations which had influential eugenics movements in the twentieth century. The misuse of genetic terms should always be a concern—as a general rule abuses follow once language is decoupled from facts—and should be vigorously rebutted by geneticists.

Genetics and Society

There may be a human tendency to feel that somehow, although technology has moved pretty fast recently, progress is reaching some sort of plateau. The opposite is of course true: technology will continue to advance exponentially for the foreseeable future and for better or worse genetics is likely to be one of the fastest moving areas. Although there are good reasons to be sceptical regarding the current value of whole genome sequencing in mainstream medicine, there will undoubtedly be advances that are not easily predictable in the light of our present knowledge. It will soon become apparent whether there are indeed intrinsic limitations to the utility of genome-scale sequencing in the context of common disease. As the potential scope of genetic testing increases and the language of genetics becomes more pervasive, it becomes proportionately more important that those outside the profession have a sufficient grasp of the main concepts to allow society to determine how the practice of genetic testing should develop. A wider understanding of the benefits and limitations of the field is needed to ensure that genetics is used in ways that contribute to human flourishing, in the broadest possible sense.

---

[6] A lack of understanding of the genetics of different ethnic groups was highlighted by the suggestion during the COVID-19 pandemic that there might be a genetic factor involved in the high morbidity and mortality affecting UK citizens in so-called BAME (Black, Asian and minority ethnic) groups—ignoring the fact that this term combines in a single category individuals from all round the world, with no more in common genetically with each other than with any other part of humanity. This misuse of genetic language does potentially matter, to the extent that it allows those in power to ignore the real contributing factors of poverty, crowded living conditions, high-risk occupations and comorbidities, factors which themselves are often directly related to historical racism.

Printed in the United States
by Baker & Taylor Publisher Services